RUSSIAN AGRICULTURE

BELL'S ADVANCED ECONOMIC GEOGRAPHIES

General Editor

PROFESSOR R. O. BUCHANAN
M.A.(N.Z.), B.Sc.(Econ.), Ph.D.(London)
Professor Emeritus, University of London

A. Systematic Studies

PLANTATION AGRICULTURE
P. P. Courtenay, B.A., Ph.D.

NEW ENGLAND: A STUDY IN INDUSTRIAL ADJUSTMENT
R. C. Estall, B.Sc.(Econ.), Ph.D.

GREATER LONDON: AN INDUSTRIAL GEOGRAPHY
J. E. Martin, B.Sc.(Econ.), Ph.D.

GEOGRAPHY AND ECONOMICS
Michael Chisholm, M.A.

AGRICULTURAL GEOGRAPHY
Leslie Symons, B.Sc.(Econ.), Ph.D.

REGIONAL ANALYSIS AND ECONOMIC GEOGRAPHY
John N. H. Britton, M.A., Ph.D.

THE FISHERIES OF EUROPE: AN ECONOMIC GEOGRAPHY
James R. Coull, M.A., Ph.D.

A GEOGRAPHY OF TRADE AND DEVELOPMENT IN MALAYA
P. P. Courtenay, B.A., Ph.D.

B. Regional Studies

AN ECONOMIC GEOGRAPHY OF EAST AFRICA
A. M. O'Connor, B.A., Ph.D.

AN ECONOMIC GEOGRAPHY OF WEST AFRICA
H. P. White, M.A., & M. B. Gleave, M.A.

YUGOSLAVIA: PATTERNS OF ECONOMIC ACTIVITY
F. E. Ian Hamilton, B.Sc.(Econ.), Ph.D.

RUSSIAN AGRICULTURE: A GEOGRAPHIC SURVEY
Leslie Symons, B.Sc.(Econ.), Ph.D.

AN AGRICULTURAL GEOGRAPHY OF GREAT BRITAIN
J. T. Coppock, M.A., Ph.D.

AN HISTORICAL INTRODUCTION TO THE ECONOMIC GEOGRAPHY OF GREAT BRITAIN
Wilfred Smith, M.A.

THE BRITISH IRON & STEEL SHEET INDUSTRY SINCE 1840
Kenneth Warren, M.A., Ph.D.

RUSSIAN AGRICULTURE

A GEOGRAPHIC SURVEY

LESLIE SYMONS, B.Sc. (Econ.), Ph.D.

Senior Lecturer in the Geography of Russia,
Department of Geography
and
Centre of Russian and East European Studies,
University College of Swansea

LONDON

G. BELL & SONS, LTD

1972

G. BELL AND SONS, LTD
York House, Portugal Street
London, W.C.2

ISBN 0 7135 1627 5

Printed in Great Britain by
NEILL & CO. LTD., EDINBURGH

To
My Wife,
Alison and Jennifer

Contents

Maps and Diagrams

Tables

Acknowledgments

This book came to be written as a result of the award of a Simon Senior Research Fellowship at the University of Manchester which enabled me to concentrate on research in 1967–68, including extensive travel in Siberia and Soviet Central Asia as well as in European Russia. The assistance rendered by the Simon Committee, Professor P. Crowe and members of the staff of the School of Geography and the Computer Centre of the University of Manchester was essential in the early stages of the work. The assistance of Professor J. T. Coppock and Mr. J. Hotson of the Department of Geography in the University of Edinburgh, in supplying information which helped in the compiling of the Atlas computer programme, saved much trial and error in automated cartography. I am indebted to Professor R. Mellor of the Department of Geography in the University of Aberdeen for assistance with sources. Help in discussion and processing statistics came from staff and students at the University of Canterbury, Christchurch, New Zealand, in 1969–70 and acknowledgment is particularly made of valuable comment by Mr. D. Greenland on the section on the climate of the U.S.S.R. Mr. P. Hanson of the Centre for Russian and East European Studies of the University of Birmingham kindly read and commented on a draft of the treatment of the economic and political background. In the final stages of the work discussion with colleagues of the Centre of Russian and East European Studies of the University College of Swansea has been invaluable. Professor R. O. Buchanan read the whole of the manuscript and offered extensive editorial help and guidance. Professor W. G. V. Balchin generously made available the resources of the Department of Geography at Swansea. For cartographical work I am indebted to Mr. G. E. Lewis and Mr. John Macdonald, respectively of the Departments of Geography at Swansea and Christchurch. The typing of the final script was carried out by Mrs. E. Wimmers and Miss J.

Davies, who coped admirably with all problems, including the unfamiliar Russian names and references.

Substantial parts of the descriptions of two of the Soviet farms included in Chapter 6 previously appeared in *Perspective* No. 6, 1970, published by the New Zealand Geographical Society, and the description of the third, together with Figure 7, first appeared in *Collectivised Agriculture*, 1970, Hicks Smith and Sons Ltd., Wellington, and permission to reproduce these items is gratefully acknowledged. All maps have been constructed from Soviet statistical handbooks unless otherwise shown.

It is impossible to name all the people in the Soviet Union who made the fieldwork possible. They include members of University faculties, chairmen of collective farms, directors of state farms, agronomists, farm workers, Intourist guides and many others. Their efforts have helped to spread understanding of the Soviet way of life not only to me, but indirectly through the impressions I have been able to use in this book and in other publications and lectures in several countries.

Finally my wife has helped extensively with research, checking and proof-reading and has undertaken the preparation of the index. If errors have slipped through, however, the responsibility must be mine alone.

University College of Swansea L.J.S.
27th July, 1971

Introduction

One-sixth of the land surface of the world and about one-seventh of its agricultural lands lie within the boundaries of the Soviet Union. The geography of this vast country has attracted many writers and research workers. The great bulk of the literature is in Russian and accordingly not accessible to those who do not read the language, but the number of books on the Soviet Union in English has grown rapidly in recent years. There is, however, as yet, a marked lack of books on particular aspects of Soviet landscapes and their economic life, and few books in English attempt a general survey of agriculture in the Soviet Union.

Yet on this vast stage has been played out one of the greatest of all dramas in social and economic organization of rural life—the collectivization and socialization of agriculture. This was the first large-scale experiment in state-controlled farming in the world. Its principles have been applied in modified form in a number of other countries where socialist or communist forms of government hold sway, notably in China, so that a third or more of the world's population is now dependent for its food on collective and state farms.

The importance to the Soviet Union of its agriculture can hardly be overstated. Since the Bolshevik Revolution, all Soviet governments have been concerned to maintain their economic advance with minimum dependence on products from other countries, though increasingly in recent years they have sought to develop normal trading relationships—so long as these do not result in strategic dependence on any other country. Home control of the bulk of food supplies and raw materials has been considered essential. For the Soviet people, an improved standard of living, in the absence of substantial imports from abroad, has required increases in agricultural output more than those needed to offset the growth of population, and because of the low standards of existence formerly common in Russia, a rise to levels even remotely comparable to those enjoyed in western countries has called for a much greater advance in agriculture.

Although Soviet practice has been to emphasize the development of industry and to derive capital for industrial development from farming, agriculture remains important in the internal economy of the country. About one-third of the national income of the U.S.S.R. is derived from agriculture[1] and some 30 million people are directly employed by farms or engaged in caring for private plots, not counting dependents.[2]

This book, the product of a geographer's approach, is concerned with the development and present characteristics of the man-land relationship. Geography is the study of the living landscape—of the interaction of human organization and environment and the effect of space relationships. Hence, the creation and development of a particular organizational type of farming in relation to its physical and social background commends itself to geographical study. Geography is, however, one of the broadest of the sciences and a geographical study is necessarily also a historical and economic survey.

A substantial amount of space is devoted to the physical environment, which in most commentaries on Soviet agriculture has been 'taken as read', though in all probability few of their readers would have had more than the most sketchy knowledge of the physical conditions in the U.S.S.R. The remedying of this omission has not been undertaken in order to excuse the poor performance of Soviet agriculture in output, but simply because it appears to the writer impossible to consider this performance objectively without some knowledge of the severity of the problems. Criticism itself is ultimately more valuable if well informed and to be effective any assessment of the worth of the collective system should have full regard for all the conditions, ranging from the pedological to the sociological, in which it was conceived and has been continued.

Understanding of the physical conditions is also essential for any attempt to predict the likelihood of the Soviet Union exporting substantial amounts of agricultural produce, as has been indicated to be at least a possibility. Such predictions are

[1] This estimate for 1965, by G. Gaponenko, head of the agricultural section of GOSPLAN, is above official estimates which attribute to industry the turnover tax yield from processing agricultural products. See Bornstein (1969), 4.

[2] *N.kh.SSSR 1968*, 446.

beyond the scope of this book, but some of the material necessary for forecasting is included, with indication of more detailed sources.

Preceding the physical section, there are brief accounts of the organizational structure of the Soviet agricultural industry and of its antecedents. These, especially the historical resumé, are intended only to supply a background for those unfamiliar with Russian history and make no pretence at originality. The ensuing survey of the economic and political scene also draws widely on works published in the west but makes use of Soviet statistical material and comments on the contemporary scene from newpapers and journals.

After examining the impact of climate and the variation of soil resources, the survey turns to the human resources—the rural population and the agricultural workforce. Next, the way in which resources are combined is seen in examples of collective and state farms drawn from personal experience. This is followed by a review of the production and distribution of each of the major crops and classes of livestock in the Soviet Union. After this systematic treatment, regional variation of production is described. Finally, a review of the development of mechanization and other aspects of improvement and the growth of conservation practices in the face of chronic soil erosion leads to concluding comments on selected aspects of the changing pattern of Soviet agriculture and its continuing problems.

As this book is concerned with scientific enquiry into the spatial patterns of production, political affairs are considered only in as much as some appreciation of them is essential to understanding the nature of the organization and the impact of the Russian form of state interference in the productive process. The object is to provide a survey of the present pattern, both in terms of overall production and regional dispersion, of agriculture in the Soviet Union as objectively as possible.

The book follows, in general, the approach suggested in the author's *Agricultural Geography*,[1] a systematic study of concepts and methodology used in the discipline, and may serve to exemplify this approach. The example, however, departs from the implied model approach in certain important aspects, some

[1] Symons (1967).

of which are inherent in the nature of the study. In particular, although statistics are published in far greater detail for the U.S.S.R. as a whole, and for individual republics and regions, than is widely appreciated, there is not sufficient regional detail published for refined geographical analysis of the whole country on a uniform basis.

A further limitation attaches to fieldwork. The facilities now offered to scientists from abroad to study in the Soviet Union are incomparably greater than they were a few years ago, and the tourist with no special concessions may visit over one hundred Soviet cities and travel with a fair degree of freedom into the surrounding country and between cities. There are, however, two severe limitations. As long as there is not complete freedom to travel the whole country, with only the limited areas of actual military and other strategic establishments specially controlled, it is not possible for such units as farms and settlements to be sampled on a scientific basis. Fieldwork must therefore remain circumscribed and actual observations confined to the areas to which access is permitted. Secondly, there is the limitation on fieldwork imposed by considerations of time and cost. No one person can hope, in the time available for a practicable research programme, to see more than a tiny fraction of so vast an area as the U.S.S.R., while travel both to and within the Soviet Union is expensive for residents of other countries. The fieldwork element in this study was limited to several short periods in the U.S.S.R., in the course of which visits were made to a number of places in European Russia, Siberia and Central Asia.

A third limitation on methodology arises from the absence of large-scale maps and air photographs of the Soviet Union, such as workers in most of the economically-advanced countries are used to having at their disposal. Few maps larger than atlas-type coverage of republics or wall-scale maps of the U.S.S.R. as a whole are available to foreigners and there are no published vertical air photographs, and few obliques, except such as appear as illustrations in books and journals.

'Agriculture' means, literally, the cultivation of fields. It is usual to understand this meaning as embracing pastoralism, in which there is some 'cultivation' of the land through care of the sward for grazing purposes. Strictly, hunting and collecting economies, in which no conscious attempt is made to improve the land, are not agricultural. There is, of course, in practice, some difficulty in classification of some cultures, and some blurring must be expected at the margins. However, in the Soviet Union, it may be accepted that in all areas in which the land is put to some use there will be conscious attempts to improve productivity, and all regions are accordingly discussed in this book. No specific attention is, however, directed to hunting, which is still an occupation of some importance in the U.S.S.R.

Forestry is referred to only in connection with conservation and the importance of its role in land use. The forest industries of the Soviet Union could not, irrespective of definitions, be treated other than summarily in a book of this length which aims at examining the whole field of agriculture, but forestry and farming are closely allied and even interdependent in most regions of the U.S.S.R.

A prime objective in geographical studies is areal differentiation—the recognition of areas with distinctive characteristics which can be distinguished from neighbouring areas. In a country as vast as the Soviet Union, such regionalization has obvious practical value in isolating the problems peculiar to particular areas and directing attention to similarities and dissimilarities of different areas. Delimitation and assessment of regional characteristics in agriculture is, therefore, a foremost part of the work of agricultural geographers and a substantial section of this book is devoted to description under regional headings. Agricultural regions should be defined in agricultural terms—associations of crops and livestock, or dominance of particular types of farming. Considerable problems of definition and methodology are involved and because of the aforementioned handicaps it is extremely difficult for a person who has not access to detailed data to produce a regionalization scheme more refined than those already in use. In general, regions recognized and defined by contemporary Soviet geographers are used in this book, minor modifications being made as appropriate.

NOTE ON TRANSLITERATION AND RENDERING OF NAMES

The system of transliteration used is that recommended by the Board on Geographical Names (U.S.B.G.N.) and is used throughout, not merely for place names. Some difficulties occur, however, in quoting translations. Thus, authors and titles of works that are available in translation are in some cases quoted in the system adopted by the translator in order to facilitate reference. Where reference is made to the original Russian source, however, transliteration follows the system here adopted.

Where a work is available in an English translation, this has normally been used and page numbers are to this edition. This is shown in the footnote as, for example, 'Milovanov (1960) 1964, 2–3', the date in brackets being the date of original publication (as for all works cited), the second date and page references referring to the translation. Some inconsistency has been tolerated in the listing of original sources in the bibliography because of the space which would have been required if all works were listed by full title and translation, but full details have been given where this appeared desirable, as where a title to a translation differs markedly from the original.

For simplicity, soft and hard signs have been omitted in the text except where a Russian term is given in italics. They are used in names and titles in the bibliography. Normally, use of Russian terms follows English usage, for example in the plural form, 'oblasts' rather than '*oblasti*'. Place names are given in direct transliteration from Russian, except for the omission of soft and hard signs and special cases such as Moscow and Georgia, and regional names such as Transcaucasia, where English forms are in general use.

Abbreviations follow English or Russian form according to whether they signify translated or transliterated forms, e.g. U.S.S.R., MTS. Capital letters are used for regions defined administratively e.g. North-west, North Caucasus economic regions, while north-west Russia, north Caucasus, etc. indicate broader geographical areas.

CHAPTER 1

The Structure of Soviet Agriculture

The distribution of farms in the U.S.S.R. by types of ownership is shown approximately by the following statistics:

TABLE 1

ORGANIZATION OF FARMING, U.S.S.R., 1968[1]

	Number '000	*Total area of agricultural land (million hectares)*	*Area of arable land (million hectares)*
Collective farms	35·6	224·0	113·5
State farms	13·4	318·6	107·4
Personal private plots	—	8·0	6·5
Individual peasant farms	0·04	—	—

Sources: Narodnoye khozyaystvo SSSR v 1968 godu, 313,330.

These four forms all have legal recognition, though individual peasant farms are now insignificant. Collective farms are governed in accordance with Model Rules prescribed by the state.[2]

All land in the Soviet Union is national state property, but that occupied by a collective farm (*kolkhoz*)[3] is leased permanently by title deeds to the workers of the farm, (*kolkhozniki*), who are, in effect, shareholders.[4] All the working

[1] For more detailed and later (1969) statistics see Appendix I.

[2] The Model Rules were first established in 1930 and revised in 1935. See Gsovski (1948–49) for full text (vol. 2) and comment (vol. 1). The 1969 revision, as finally adopted, was published in *Pravda* and *Izvestiya* 30.11.69. For translations into English, see *Third All-Union Congress of Collective Farmers*, Novosti Press, Moscow, 1969, and *C.D.S.P.* 21 (50), 9–15.

[3] *Kolkhoz* is an abbreviation of *kollectivnoye khozyaystvo* (collective farm).

[4] Until recently the association of members of a kolkhoz was referred to as an *artel'*—a term in medieval Russia for co-operatives in fishing, building, manufacturing etc.—but in the revised Model Rules (1969) the word 'artel' was omitted on the grounds that it is not appropriate to the large, modern collective farm. *Pravda*, 26.11.69.

capital of the farm belongs to the kolkhoz except for dwellings and the small tools and livestock privately owned in connection with the garden plots which members are permitted for their own use.

A general meeting open to all the members of the farm is, in theory, the highest managerial body of the farm, and must be convened not less than four times per year. As well as discussing policy and a variety of matters, such a meeting elects the farm board, which is the executive body of the farm, its auditing commission and a chairman, each for three years. The chairman holds a powerful position and must be acceptable to the Communist Party. He is responsible for the running of the farm from day to day with the aid of the farm board, which meets not less than once a month, and the specialists, who may be hired if not available from within the farm.[1]

Labour is divided into sections, departments, brigades, teams or other production units according to the needs of the individual farm, and each unit has land, machines, implements and buildings assigned to it. Each unit is expected to maintain proper accounts and the head of the unit is responsible to the farm chairman and board for its proper functioning.[2]

Membership of a kolkhoz is hereditary, young people on the farm being eligible for full membership at 16, and other citizens may be admitted by a vote at a general meeting of the farm. Similarly, an application by a member to leave the collective goes before the assembly of members and the farm board makes a financial settlement with the former member at the end of the accounting year. Expulsion may be used as an extreme measure against a member who systematically violates labour discipline or the rules of the farm, subject to appeal to the local soviet. Normally, however, such behaviour is dealt with by the farm board or assembly by one of a range of penalties including censure, severe reprimand or transfer to lower paid work.[3]

Leave of absence may be given to members who are on active military service, elected to public offices, on full-time

[1] Model Rules (1969) Nos. 46–52 and 24.
[2] Model Rules (1969) Nos. 26 and 53.
[3] Model Rules (1969) Nos. 3, 7 and 35.

study or working in intercollective farm organizations or other approved bodies, and membership of the elderly and disabled is continued as long as they reside on the farm.[1]

Farm policy and production are theoretically decided by general meetings of members of the farm, but in practice the national plan requirements and its local application severely limit the farm's freedom, and one of the chairman's main tasks is to try to ensure that the farm fulfils the quotas of commodities required for the plan.[2]

Compulsory deliveries to the state are paid for at prices fixed by the state and altered from time to time. Prices for some commodities are varied regionally to compensate for the differing costs of production resulting from geographical factors. The surplus not required by the state can be sold at market prices, which are usually higher but controlled. The remuneration of the farm workers depends on total proceeds, less operating expenses and taxes. The deliveries required by the state are specified in absolute terms, so farms well endowed with land and other factors of production can remunerate their workers better than those which are less fortunate.

The basis of calculation of the individual's earnings is the labour-day unit.[3] Different kinds of work are evaluated differently and the worker in a skilled job in a highly productive section of the farm can, therefore, earn much more than the average, but thc kolkhoz is expected to fix guaranteed minimum earnings.[4] Cash payments used to be small, much of the remuneration being in animal fodder and other payments in kind. In recent years, however, there has been a considerable improvement, and the requirement for farms to make monthly cash payments is written into the Model Rules.[5] Until recently, the collectives were entirely responsible for any pension systems that they chose to operate, but the state now guarantees pensions for retired kolkhozniki. The amount

[1] Model Rules (1969) No. 6.

[2] Compare Venzher's definition of a collective, quoted p. 80.

[3] The actual term *trudoden'* is not now favoured, the system of basic time and piece-work rates and bonuses being very much more complex than the labour-day concept suggests (see Model Rules (1969) No. 27).

[4] Model Rules (1969) Nos. 4 and 28.

[5] Model Rules (1969) No. 30.

of a pension is related to the individual's earnings record, but a kolkhoz can add to the minimum pension.[1]

The most common type of state farm is the *sovkhoz*,[2] which is normally producing for the market, though it may have research interests as, indeed, may a kolkhoz. In addition, there are other state farms which are essentially research and experimental establishments. State farms are financed by the government, and until recently the state required them to deliver their produce at prices lower than those paid to collectives and, in return, subsidized farms that failed to make a profit. In 1967 it was announced that sovkhozes were to be transferred gradually to a self-financing basis and would receive the same prices as collectives for their produce.[3] They were to be free to plan their own production within certain limits, to vary their labour force and to dispose freely of produce not required by the state procurement agencies.[4]

Wages are paid mainly in cash, the rates varying with the work performed and the qualifications and success of the individual. State farms are generally larger and better endowed than collective farms, although there are exceptions to this generalization and the recent reforms are narrowing the gaps between the two types of farm. Reference will be made later to the growth in importance of the state farms at the expense of collectives, particularly in the past decade, and to wages and other conditions of rural employment.

PERSONAL PLOTS[5]

Since the early days of collectivization, members of collective farms have been allowed to hold small areas of land for personal and family subsistence. Similar facilities have been extended to certain wage-earners and retired persons, but with variations from time to time and in different areas.

On collective farms, personal plots are allotted primarily to members working on the farm, but they may be retained by the families of those who work or study elsewhere with the

[1] Model Rules (1969) Nos. 39 and 40.
[2] Abbreviation for *sovetskoye khozyaystvo*.
[3] *Pravda*, 15.4.67.
[4] For comments, see Clarke (1969), 162–165.
[5] The Russian term is *priusadebnyy uchastok*.

consent of the farm and when only minors remain at home. Plots are also kept on retirement and by the families of deceased members. Teachers and other employees residing on the farm may be allowed plots but no plots may be alienated or worked with hired labour. Plots vary in size between 0·15 and 0·50 hectares (0·37–1·24 acres) according to local conditions. On irrigated land, however, they may not exceed 0·2 hectares. The general membership meeting may vary plot size according to the size of the family and its participation in communal work. Plots are maintained as units, only movable assets being available for distribution in the event of division of a household. Individuals may also own livestock, which may be pastured on the collective lands. Normal permissible holdings for a family are a cow and calves, one heifer or steer up to 2 years old, a sow and piglets or two pigs for fattening, ten sheep and goats, unspecified numbers of poultry and rabbits and twenty beehives.[1] In nomadic regions, personal holdings may rise to 8–10 cows, 100–150 sheep and goats, 10 horses and 5–8 camels with young.[2] The collective gives assistance with acquiring livestock, provision of fodder, transport and specialist services.[3]

Workers on state farms and other employees may hold personal plots subject, in the Russian republic, to the following limitations:

TABLE 2

PERSONAL PLOTS PERMITTED IN THE R.S.F.S.R. TO STATE WORKERS

	hectares
State farm workers	0·15–0·50
Forestry workers	0·50–0·75
Rural teachers, transport workers, agricultural specialists, etc.	Max. 0·25

Source: C.D.S.P., 13 (21), 3–5 (from *Sovetskoye gosudarstvo i pravo* No. 5, 1961).

In practice, in every republic, plots for state farm workers are usually smaller than those of members of collective farms.

[1] Model Rules (1969) Nos. 42–44.
[2] Standard Charter (1935), trans. in Hubbard (1939), 135, and Gsovski (1948–49) vol. 2, 445–446.
[3] Model Rules (1969) No. 43.

On some state farms the keeping of personal cows is discouraged, if not prohibited. On state farms, too, the housing commonly consists of blocks of flats, in which case plots are separate small allotments, whereas on collective farms they are usually large gardens attached to the houses. On collectives also, if the residential area permits only small gardens, additional lands are allocated in allotment areas, subject to the total for any one family not exceeding the stipulated maximum area.[1] Collective farmers are given assistance by the collective to build their own houses which may be sold to other members if they leave the kolkhoz.

The value of the personal plots and the fluctuations in official attitudes towards them are discussed later.

So far, only those organizational forms that have general applicability and statutory recognition have been described. Reference must also be made to the *zveno* system of organization of labour within the large farms. The zveno or 'link' is a team of peasants (usually 3 to 8 persons) who are allowed to operate an area of land belonging to a collective or state farm, using the facilities of the farm for machinery and inputs as well as specialist advice, and deriving their income from the output they produce for the farm. It appears to have been first used in the 'thirties for labour-intensive crops such as sugar beet and cotton.[2]

The idea of the small semi-autonomous unit working within the framework of the collectivized economy was given a measure of official blessing during and after the Second World War, apparently to ease the restitution of the collective system after it had been weakened during the war, especially in the areas occupied by the Germans.[3] In 1948 a government decree ordered specific plots to be attached to each brigade or zveno to encourage interest in productivity, but there was a violent change of policy in 1950 and the advocates of the zveno came under official attack. Under Khrushchev's rule the links began to emerge again and after the bad harvest of 1965 several state farms came out in support of them and a conference was held to discuss the advantages of the link

[1] Model Rules (1969) No. 42.
[2] Volin (1951), 29.
[3] Pospielovsky (1970), 413.

system.[1] Much evidence was produced in support of the links and from 1966 official backing grew slowly. Conservative officials continued to oppose links, partly because of the abnormally high earnings obtained by teams which were favoured in supply of machines and other inputs.

Many of the links were formed to deal with only one crop rather than to care for a piece of land on a long-term basis and many of the potential advantages of the system were thereby lost. Large numbers of links were disbanded for various reasons. Great variety has been apparent in the organization of links, with areas from under 100 to over 3,000 hectares being allocated to one team and greatly varying degrees of control by the parent farm.

The revised Model Rules make provision for links and other production subdivisions of a kolkhoz with internal accounting (*khozraschet*)[2] but a provision included in the draft for allocation of land to such a unit for a number of years was omitted in the final version.

An alternative to the zveno, with apparently little effective difference, which has been adopted by some farms, has been to reduce the size of brigades to 5–15 members and make these units responsible for their own accounts.

INTERCOLLECTIVE ORGANIZATIONS

Collective farms may join forces to undertake development projects beyond their individual resources or which need to serve an area larger than one farm. Intercollective farm organizations for construction work have been active since the late 'fifties and in 1967 numbered 1,170, distributed through 58 oblasts, krays and A.S.S.R.'s of the Russian republic.[3] In addition to construction they included design institutes, lumbering units and other enterprises. Intercollective organization is favoured for constructing and operating plants to supply materials to farms and to process surplus produce where the state facilities are inadequate.

As an example of the former, some Crimean farms united forces in 1960 to produce cement from local minerals. In 1966 production was 360,000 tons and the plant supplied

[1] Pospielovsky (1970), 418ff. [2] Model Rules (1969), No. 26.
[3] For types of administrative divisions, see Appendix II

state farms and other users as well as the shareholding farms.[1] Such enterprises clearly meet a real need, though with two obvious disadvantages: any plant built from even the pooled resources of several farms will be relatively small and will probably operate at a high cost compared with a modern state plant. At this cement plant, costs were assessed at 11 rubles 32 kopeks per ton compared with about six rubles per ton in large plants. It appeared, however, that the retail price of cement was 36 rubles per ton, so the intercollective plant had ample operating margins and would become unprofitable only if the price of cement was brought down with rising production. A more likely danger for such an enterprise is the possibility that a state plant will be built to meet local needs more fully, entailing the closure of the intercollective plant. This happened to many intercollective hydro-electric installations when the electricity grid reached their localities. The balance of advantage, however, seems to lie with the establishment of needed plant, without which many years of shortages may have to be endured.

A major field for intercollective enterprises is the processing of fruit, vegetables, meat and other farm produce. Reports of plants being established or being needed occur frequently in the Soviet press. Farms may also undertake auxiliary enterprises to produce items of which their own needs form only a small part of the marketable production. There are, however, risks in this diversification. One farm which was turned from a poor to a highly profitable unit by manufacturing cable terminals, switches, etc., came under attack and the chairman was prosecuted, but after a hearing the charges were dropped.[2]

Provision is now made in the Model Rules for both intercollective organizations and kolkhoz-sovkhoz joint enterprises.[3] Auxiliary industrial undertakings are also recognized and a kolkhoz may contract with industrial and trading organizations for workshops to be set up.[4] They must not, however, supplant agriculture or be detrimental to it in the communal economy.

A variety in intercollective farm organizations initiated in Leningrad oblast in 1969 took the form of a combine formed by

[1] *Sel'skaya zhizn'*, 30.5.67, 2; *C.D.S.P.* 19 (23), 33.
[2] *Izvestiya*, 4.8.67 and 8.6.67; *C.D.S.P.* 19 (23), 17–22.
[3] Model Rules (1969) No. 18.
[4] Model Rules (1969) No. 17.

nine collective farms. They planned for coordination of production through an intercollective farm council elected by the representatives of all the farms involved for two years. The council operated through an executive board and employed specialists and fixed wage and incentive rates.[1]

ADMINISTRATIVE CONTROL

Any attempt to describe the organizational machine controlling agriculture faces the fact that it will probably be out of date before it is published. Thus, since the death of Stalin, several major reorganizations have been announced, notably in 1953, 1955, 1958, 1961, 1962–1963, 1964, 1965 and 1966. Nevertheless, some features remain little altered and there is some pattern in the alterations.

Clearly much time and effort have been involved in making changes which have not endured, and to this extent there has been considerable organizational waste. Nevertheless, this constant flux has some advantages, as Swearer has pointed out:

> It breaks up interlocking family groups and mutual protection societies which short-circuit central control channels. It permits the infusion of new faces and ideas into administrative ranks.[2]

The personnel, however, do not always change when the system changes, otherwise there would be chaos. There is a considerable element of continuity in a post although the designation of the official may be changed, and, of course, some of the administrative staff are constantly being replaced for various reasons. Major changes necessarily took place during the period when Khrushchev was attacking the excessively centralized apparatus maintained by Stalin.

Khrushchev's organizational changes did not always have the desired results. His attempts at interference with the bureaucracy sometimes made matters worse because of confusion created in the agricultural administration, and some of the effects of decentralization were so unsatisfactory as to require a retreat in the direction of renewed control from the centre.

The general trend over the past fifteen years, however, has

[1] *Pravda*, 9.10.69, 2; *C.D.S.P.* 21 (41) 20.
[2] Swearer (1963), 10.

been to give farm chairmen and managing committees more freedom to determine the pattern of output they consider most suitable for their own physical and economic conditions, within the general production pattern laid down by the central and regional committees. From livestock farms visited, for example, it appears that they will be required to concentrate on, say, milk production, and the regional committee may stipulate main subsidiary enterprises, perhaps beef and pig production, but the farms seem free to decide, within the availability of the necessary inputs, whether they also, for example, keep sheep, which crops they grow for fodder and how they arrange their field rotations. In short, farms appear free to produce whatever crops or livestock they wish once they have made the basic allocations for a reasonable attempt to fulfil state plan requirements.

This applies particularly to the collective farms. With state farms, though the management must have reasonable operating freedom, closer control is to be expected.

As with all forms of production of national importance, the Communist Party is the ultimate controlling power, working through its own regional and local bodies as well as through the offices of the government.

The general plan from the U.S.S.R. State Planning Committee (GOSPLAN) is handed down to the oblast planning organization and thence to the rayon level, from which the collective and state farms receive details of the commodities they are required to sell to the state and the amounts needed to fulfil their individual farm plan. The Ministry of Agriculture of the U.S.S.R. is extremely powerful. It gives instructions and decisions to the Ministry of Agriculture of the R.S.F.S.R., accommodated in the same building in Moscow, and the other Union Ministries.

The chain of command thus follows the ordinary administrative structure. The Territorial Production Administration set up by Khrushchev shortly before his removal from power had cut across this hierarchy but his successors promptly abolished the TPA and restored the rayon as the basic local unit of agricultural administration.[1]

[1] Clarke (1969), 159.

CHAPTER 2

The Evolution of the Agricultural Pattern

The pattern of agriculture in the area that now comprises the Soviet Union has been evolved over thousands of years, with development in the various parts of the widely separated territories following different courses. Though in the present stage of collectivized and state-directed farming the objectives of different regions are largely determined by decisions made in Moscow, the practical and economic responses to physical conditions which have influenced man's use of the soil from the earliest times are necessarily preserved.

The early human inhabitants of the earth were probably, of necessity, nomadic in their habits. The distances travelled by most individuals in their generally short lifetimes were small, but in central Asia, including parts of the present Soviet Union, there arose the great tribes of horsemen whose migrations over thousands of miles across the steppes may have been accomplished in relatively short periods. These tribesmen, however, were not true agriculturalists tending the soil, but rather pastoralists exploiting the resources of the natural grasslands, and it was probably at least partly the search for fresh pastures that drove them to move frequently.

The areas believed to have provided the hearths for the origins of true sedentary agriculture are to the south of Soviet territories. The earliest use of simple cultivation techniques may have occurred in many favourable areas from China to Africa or even in the Americas[1] but archaeological evidence indicates that mixed farming, in which grain growing was combined with livestock rearing to provide a prototype for modern methods of mixed farming in temperate regions, probably evolved in the eastern Mediterranean area, extending

[1] Vavilov (1935) 1949–50; Sauer (1952).

to Persia and India. The valleys of the central Asian ranges, offering sites protected from the cold air masses of the north, streams supplied in the warm season from melting snows, and light soils, easily turned with the digging stick or hoe, may well have attracted permanent settlement at very early dates.

According to Vavilov, (who, with his Russian colleagues, identified eight main centres of independent origins of the world's cultivated plants in both the Old and the New Worlds), central Asia, including Tadzhikistan, Uzbekistan, and the western Tyan-Shan ranges as well as Afghanistan and north-west India, provided over forty species for cultivators. These included a range of wheats, important legumes including peas, lentils and beans, and cotton.[1] Still richer in their endemic species were considered to be the uplands and valleys of Transcaucasia, Asia Minor, Iran and Turkmenistan. There, Vavilov found evidence of the origins of 83 species, including grapes, pears, cherries, figs, walnuts, almonds, lucerne and different species of wheat and rye.

When Palaeolithic men advanced into the northern forests of Europe, after the waning of the last continental ice sheets, they possessed no such crops, nor the facilities for settlement based on any economy but hunting. The range of animals was, however, considerable, including the mammoth, woolly rhinoceros, cave lions and wild horses and oxen, with reindeer and seal further north. Fish would be a valuable source of food along the rivers, which provided the main routes of travel, as well as in coastal areas.

Through this and the succeeding Mesolithic period, the soil and vegetation belts we know today were becoming established, and the rudiments of cultivation were being carried into southern Europe from the Near East. The method of cultivation was that of shifting or forest-fallow exploitation. The natural or regrown vegetation was burnt to clear it and fertilize the ground with the ashes, and seeds sown and the soils turned over with hoes or plough-sticks. After several crops the land would be allowed to revert to rough pasture or scrub and fresh clearings were made. Easily worked soils were naturally preferred, such being the forest soils and the modified chernozems of the forest-steppe area. Crops included the primitive wheats, barley, beans,

[1] Vavilov (1935) 1949–50.

peas, lentils and flax. Livestock included oxen, sheep and pigs, but numbers would not be sufficient to offset a shortage of manure, which in itself would have forced cropping to remain of a shifting nature in forest clearings. Riverside meadows gained in importance as livestock numbers increased. In the Neolithic period, variously dated at 5,000 or 4,000 to 3,000 or 2,000 B.C., cultural landscapes of the north and south of European Russia, Central Asia and Siberia were becoming more clearly differentiated. The cultivators and livestock herders of the forest-steppe and steppe were developing much more rapidly than the peoples of the northern forests and were probably five hundred years ahead.[1]

TRIPOL'YAN CULTIVATORS & NOMADIC INVADERS

During the period 3,000 to 2,000 B.C. the Bronze Age culture of the lands between the Dnepr and the Pruth, known as *Tripol'yan*, had adopted the use of metal tools, developed originally in the Carpathian and Caucasian regions, which were rich in copper. The light plough, the ard, facilitated cultivation, which remained, however, on a slash-and-burn basis, with areas in which cultivation was rotated, rather than defined fields.[2] Wheat and millet were important and hay was cut and conserved for the livestock. Villages became permanently established, with up to two hundred dwellings, and houses of timber and clay were often quite substantial and divided into several parts. Equipment included stoves for heating and querns for grinding grain as well as a growing range of artifacts of metal and pottery.

The agricultural practices of the dwellers of the western wooded steppe and steppe regions suffered in the early Iron Age from the incursions of the Scythians. These were the first of the nomadic peoples who invaded and occupied the steppes, imposing their rule on the agricultural dwellers and exacting tribute from them. Initially they were wild and destructive but the near proximity of the Black Sea steppes, in which they settled, to the Greek colonies led to the development of a Scythian civilization which endured until about the 2nd century B.C. As the Scythian power was declining, that of the

[1] Rybakov (ed.) (1966), 41.
[2] Smith (1959), 63.

Sarmatians, a people described by Herodotus as dwelling in the Don region, was increasing. They appear to have occupied Scythian territory and to have ruled over the steppes and dominated regional trade for the next four hundred years while agriculture evolved slowly.

During the later periods of the Scythians and the Sarmatians the more sedentary people were building the foundations of what were to become the Slav princedoms. Trade routes following the Dnepr, Don and Volga river systems facilitated interchange not only of products but of ideas and techniques, so that primitive agricultural methods and improved grains and livestock penetrated into the forest zone during the Iron Age. The number of Slav people in the forest and forest-steppe zone appears to have grown by migration as well as by natural increase, and they gained in strength in the agriculturally favourable western steppe area between the Pruth and the Dnepr. The evolution of the Slavs into political units was hindered by invasions of their areas of settlement by the Goths in the third century A.D. and the Huns in the fourth century. The Hun victories over the Goths, however, opened the way to more effective colonizing by the Slavs.

After the Huns came the Avars and then the Khazars in the seventh century, and in each case the descendents of these terrorizing nomads settled and adopted a way of life based on farming and commerce. Many other peoples lived in the forest-steppe, the forest clearings and the river valleys—Volynians, Ulichians, Polyanians, Vyatichans and others—some of their homelands being still identifiable in the place-names of European Russia. The more southerly of these communities had the advantage of working the rich chernozems in a kinder climate but the disadvantage of exposure to the repeated waves of invaders from across the steppes. The more northerly groups, such as the Drevlyans of the Pripyat marshes, were better protected against invasion but had much inferior land. Agriculture was carried on in clearings in the forest, planted with crops for a few years and then abandoned, but hunting, fishing and bee-keeping were probably the main means of support for these people of the poorer lands, with furs, honey and beeswax the principal items they traded along the rivers from the Baltic to the Black Sea.

Along the river routes from the north came the Vikings or Varangians, as the Slavs called them, and these Scandinavians became the most consistent raiders known in the forest zone, yet they also supplied settlers, who intermarried with the Slavs and other peoples of the forest and steppe. In the ninth century they controlled most of the country from the Baltic Sea to the middle Dnepr, and by 862 Rurik had firmly established Novgorod as the capital of a state which was still vaguely defined but had become known as Rus. In 880 his successor, Oleg, captured Smolensk and, two years later, Kiev. Other embryos of the Russian state were placed by the Varangians, supported by Slav and other followers, by the Sea of Azov and on the lower Volga to control the trade routes, but there was to be no peaceful evolution in these steppeland thoroughfares. When the Khazar grip on the steppes was destroyed, the Pechenegs came to raid Kiev, but Kiev survived to become the principal city through which Christianity entered the Russian lands, and in its wake came contacts between its princes and royal houses throughout central Europe, to be solemnized in marriage bonds and capitalized in commerce.

In Kievan Rus there was probably some development among the ruling classes of more advanced forms of agriculture with permanent fields. 'Slash-and-burn' techniques, however, remained widely in use for clearing land to be cropped for a few years at a time.[1] Millet appears to have been the main crop, its drought-resistant qualities proving valuable in forest-steppe conditions. Cultivation was aided by a variety of implements. Archaeological investigations have revealed ard irons, plough shares and coulters in the wooded steppe settlements dating from the eleventh to thirteenth centuries, while irons from the light *sokha* ploughs have been found from earlier dates in the true forest area. Reaping was carried out with sickles, and short-handled scythes, threshing probably with flails and winnowing with a spade in the wind. Grain storage pits were made carefully in Kievan Rus, implying the drying of the grain.[2] Improvements were carried northward, and by the eleventh or twelfth centuries even Novgorod in the far northern tayga had its established field systems, with wheat and rye the

[1] Smith (1959), 67.
[2] Smith (1959), 110.

main cereal crops, as appropriate to the more northerly latitudes. There was increased reliance on domestic livestock, as opposed to hunting, in the forest zone by the eleventh century, but the structure of herds is a more difficult question to solve. Probably pigs were generally most numerous, followed by cattle, sheep and goats.[1]

More farming peoples moved into the steppe area during the period of Kiev's supremacy but, in the latter half of the eleventh century, the steppes again suffered from raids by the nomads. The Polovtsy or Cumans sacked Kiev, and though it recovered it did not regain fully its former glory but rather degenerated into a group of warring principalities. Refugees from Kiev and the forest-steppe area migrated into the forest zone, preferring improved security to remaining in constant danger on more fertile land. Though the forests were generally relatively infertile, patches of good soil, including that west of Vladimir, fostered a numerous farming population.[2]

In the deciduous and mixed forest zones agriculture became more intensive as the population increased. By the twelfth century, the population of the Russian lands was about 7½ million.[3] Exports from the forests included furs, honey and wax, and from the farms, flax, hemp, hides, skins, suet, tallow and grains. Local industries also made use of these commodities, so that there was pressure on available agricultural land. 'Slash-and-burn' techniques continued, exploiting the fertility accumulated under the forest cover, but fields with primitive rotations were increasingly developed. Two-field systems, with one field fallow, gradually developed into three-field systems, from which, as two fields were productive each year while only one remained fallow, total output was greater. The light sokha type of plough was generally used in the northern districts where boulder-strewn soils resulting from glaciation were common, while the cumbersome *ralo* was used on heavy soils. Heavy ploughs needed a team of draught animals which necessitated communal working and may have been one of the factors encouraging the growth of the commune and the gradual introduction of slavery and serfdom in the interest of

[1] Smith (1959), 115.
[2] Parker (1968), 56.
[3] Vernadsky (1948), 104–105.

those who were establishing themselves as landowners. The large estate (*votchina*) developed a considerable output of wide variety, including most of the products listed above, and the practice of collecting dues from the dependent peasants added to the wealth of the landowner. In addition to these estates worked by peasants who tended to become increasingly dependent, there were monastic farms and smaller peasant holdings. The wealthy townsfolk and landowners were now beginning to live well but the lot of the peasant had scarcely improved in four centuries.[1] Indeed, the thirteenth century saw deteriorating conditions for all classes as the Mongol invasion developed. From 1237 the Mongols, or Tatars as the Russians called them, dominated the forest-steppe zone and struck deep into the forests, destroying and looting and bringing all the lands of east Russia under their control. Novgorod alone retained independence but even Alexander Nevsky and his successors had to accept the Tatars as overlords. This weakening of the Russians enabled the Lithuanians to advance from the west and extend their kingdom over west Russia.

As time passed the Tatar yoke became less onerous, though assertion of independence by a Russian prince usually brought quick and vicious reprisals. The steppes had been largely abandoned by farmers and reverted to grazing lands for the nomadic and unsettled peoples. Nevertheless, firm control by the Tatars eventually stabilized trade routes through the steppe lands and the Don and the Volga emerged as the new main river routes. Moscow, relatively secure within the forest zone, benefited from this changing situation. When, in the fifteenth century, the Tatar domination of Russia was reversed, it was the Grand Dukes of Muscovy, Vasily I and II and, above all, Ivan III, who provided the leadership, so that in 1547 Ivan III could claim the title of Tsar of All the Russias.

During the occupation by the Tatars, the early lead in central European civilization given by Kiev had been utterly destroyed and when Moscow came to the fore it was in many ways an inferior state that developed. The Tatars had demonstrated the success of power, authoritarian government and repression, and the Russian tsars emulated them as they carved out a new empire. The upper classes of boyars and gentry

[1] Parker (1968), 64.

emerged as the dominant group and their great estates became centres of local power and wealth. The middle classes of merchants, artisans and independent farmers had suffered badly and had little influence. Increasingly, the tsars granted lands to their supporters, and the peasants found themselves further bound to these estates. Many were slaves, but serfdom was not yet as entrenched in Russia as it had been in western Europe in the twelfth and thirteenth centuries under the feudal system.

How serfdom became an integral and fundamental part of the social system in Russia at such a late stage is not wholly clear. It would appear that the appanage princes and non-titled landowners, the principal landlords of the fifteenth century, held their land by grants or on contract from the grand dukes of Moscow or other princes. In return, they owed service to the grand prince, who, by the early years of the sixteenth century, was effectively the Grand Prince of Moscow (unless the landowner chose to serve the rival houses of Lithuania or Poland). Ivan the Terrible weakened the boyar class with confiscation of lands which were redistributed among lesser gentry, who then owed their position to the crown. His new system of service fiefs helped not only in his military enterprises but also with colonization, but the landowners could exploit their lands only by greater subjection and exploitation of the peasantry.

The masters steadily increased their jurisdiction over their servants whose bondage became registered and hereditary. In 1649 the Code of Tsar Alexis gave landowners the right to pursue and reclaim runaway serfs. Many, nevertheless, absconded, sometimes taking up service with other landowners, sometimes wandering as vagrants or settling on the frontiers of the southern steppes, strengthening the Cossack communities which had been established there. Others went east beyond the Volga. In the Ural region employment was available in the mines, and Siberia offered all the lure of the furthest frontiers and complete escape from serfdom. Some of those who stayed behind managed to conserve their status as free, tax-paying peasants, later called 'black ploughlandmen', but this became increasingly difficult as enserfment spread. In Great Russia there developed a range of degrees of dependence, down to the slaves, who were totally the property of their masters. These slaves and those in temporary slavery, the 'referred slaves', paid no dues to the state, but dues were paid, either directly or indirectly, by the serfs and metayers, who were less than totally enslaved.

Two main forms of relationship between serf and owner evolved. Under the *obrok* system the serf made payments to the landowner in return for an allotment of land to work for himself, while under the *barshchina* arrangement the serf earned his allotment solely by labour. In either case a serf worked his owner's land. In addition to the serfs there were state peasants, who had a greater degree of freedom.

The serfs and state peasants were involved in varying degrees, according to the region, with the commune (*mir* or *obshchina*), which was responsible for various administrative, fiscal and agricultural activities. The commune had originated with the allocation of responsibilities for tax collection, law enforcement and other local duties, probably mainly during the Kievan and Mongol periods. Later it became convenient for the landlord to arrange work on his estate through the elders of the commune which also became responsible for regulating the use of meadows, forests, rivers and other resources. In the sixteenth century, scattered settlements of not more than eight or nine households were the usual form of village, with, however, a trend to greater nucleation and the more widespread adoption of the three-field system. The open fields required control of strip cropping, and periodical redistribution of strips gradually became common, at least in the central forest zone of Russia. The commune shouldered the responsibility of such redistribution, and controlled the primitive rotation systems. These were based on an autumn-sown cereal, wheat (where practicable) or rye, and a spring-sown crop, barley. Millets and buckwheat were important in the warmer and drier areas, flax and hemp in the cooler north-west. The sokha was the normal type of plough, and harrows and scythes were in fairly widespread use.

In the northern regions the holdings of land varied from private individual farms to complicated systems of joint ownership, but after much unrest in the eighteenth century the state imposed a communal land system with periodical redistribution of strips, which slowed down the emergence of a more prosperous class of peasant. In the steppes, settlement expanded slowly after the Tatar defeats, accompanied first by horse, sheep and cattle rearing and then by wheat growing. It remained sparser than in the forest-steppe and the land was largely held in great estates, with the commune less important. The gentry brought serfdom and their three-field system, however, into parts of the steppes, especially the wooded northern margins, so that there was no sharp division of agricultural types but rather a gradual transition from the forest to the chernozem soils.

By the late eighteenth century the agricultural commune and the three-course open fields with periodical redistribution of strips had become the dominant type of agricultural exploitation in Russia and in many parts of the lands into which the Great Russians had expanded, though there was much variation in detail. The commune might be based essentially on one village, or several villages, or might include part only of a large village. Its membership was hereditary, though newcomers could be admitted. Movement, however, was limited by the commune, which had an interest in keeping its own labour force intact during the long periods between revisions of the poll-tax assessment, because this was applied by Peter the Great on a population basis equally to the serfs and state peasants. The poll-tax replaced earlier direct taxation on households and still earlier assessments on ploughed land, which had been collected by the commune. The collection of the poll tax increased the role of the serf owners. As the poll tax was a fixed rate there was an incentive to increase cultivation and crops for sale. The burden of serfdom became heavier as the owners progressively exploited their labour resources to supply the growing markets in the towns and the export trade in wheat, though owners were legally required from 1734 to assist their serfs through famines and hard times.[1] Trade was stimulated by the abolition of internal tolls (1754) and restrictions on the corn trade (1762).

The peasants' burdens were not limited to taxation and the exploitation of their labour by their owners. Peter the Great used forced labour for the construction of canals, military and naval works, the reclamation of swamps and the building of St. Petersburg. He introduced the passport system, which restricted

[1] Sumner (1947), 154, 1961 ed. 135.

the legal movements of serfs to those required by the government and their owners. Catherine the Great gave further powers to the owners until they were in almost complete control over their serfs.

As previously noted, oppression was countered by the more daring and enterprising peasants by flight to join the Cossacks in the south or the colonizers and fur-seekers in Siberia. Local rebellions flared up continually, but for the most part were easily controlled because of their local nature. Exceptions were the mass revolts led by Bolotnikov (1606–07), Stenka Razin (1670–71), Bulavin (1707–08), and Pugachev (1773–75). These rebellions, amounting to civil wars, were only partly agrarian in origin, but the combination of the serfs and other peasants with the militant leadership provided by the Cossacks made these revolts of a magnitude to threaten the Moscow governments. In each case, however, Moscow prevailed, partly because of the lack of unity among the rebels, for there were differences of objectives between peasants of the countryside and townsmen, between the Moslem Tatars and Bashkirs and the Orthodox Russians and between the militant Cossacks and those who had themselves become established landowners. Pugachev, at the peak of his successes after taking Kazan, proclaimed the abolition of serfdom and the appropriation of the land for the peasants some ninety years before the Edict of Emancipation.

Rebellion, whether regional or local, probably achieved little directly towards obtaining better conditions for the peasants, because it always induced repression and the tightening of control. Nevertheless, the battles and suffering became part of the background against which the revolutionary movements of the nineteenth and twentieth centuries evolved. In time, these captured the support of the intelligentsia, who were eventually to organize the overthrow of the regime. Continual unrest and the many murders of serf-owners and their bailiffs must also have prepared at least the more enlightened landlords to recognize the need for change. It was, however, the industrial and commercial revolution that was spreading through both town and countryside in the nineteeth century that provided the conditions in which the abolition of serfdom eventually appeared as the most urgent of many essential reforms. The availability of labour for the factories was restricted by the large proportion of the potential working force bonded to landowners, and the low incomes of the great mass of the people limited the rise in demand for factory products. Even in the agricultural areas, especially where crops like wheat were grown on large estates for export, many landowners recognized the inefficiency of serf labour and were anxious to abolish the system. Redemption also offered a means of raising capital and reducing the burden of debt that afflicted many landlords. Economic advantage seemed for the entrepreneur and the commercial estate owner to be allied with the prospect of greater stability in the countryside if reform could be achieved on their terms.

Nevertheless, opposition to the proposals for liberation put forward by the government was widespread, and Alexander II wisely involved the gentry in discussion on how emancipation was to be achieved. Initial impetus was gained from a request from the Lithuanian gentry to emancipate their peasants, though Emmons states that this affair was arranged in the Ministry of the Interior 'for the explicit purpose of creating an impression that the gentry themselves were anxious to cooperate in the government's undertaking'.[1] The provincial gentry committees, set up to examine the situation in their own districts and report to the government on recommended means of achieving emancipation, reflected differing views, some of which were derived from the varying values of land and labour in their respective regions. Their work was, however, hindered by lack of clear directives from the government on the extent of their powers, and while they conducted their discussions the essentials of the programme were being worked out in St. Petersburg. Indeed, had this not been so the conservative majority among the landlords would certainly have formed a force united in its intention to obtain the maximum benefit for the landlord class from the emancipation. When the deputies from the provincial committees assembled in St. Petersburg, the government's proposals to grant the landlord the right to retain one-third of his lands aroused intense criticism, almost all arguing in favour of two-thirds, and the principle of minimum sizes for peasants' plots was attacked in an effort to whittle down the amount of land to be handed over.[2]

[1] Emmons (1968), 60.
[2] Emmons (1968), 248–249, 311.

Although their representations had little effect upon the final form of the emancipation, the gentry were at least prepared for its conditions. The peasantry expected emancipation on far better terms than they got. They expected the land to be granted to them free but found themselves burdened by redemption payments for 49 years; they expected complete freedom and found themselves still tied to the commune; they hoped for larger holdings and discovered their allotments to be generally too small for their economic independence. The landowners were mostly allowed to retain substantial areas of the best land, together with the rights of use of grazings and forests, and many of the freed serfs received allotments much smaller than the pre-reform holdings.[1] The 'beggar's allotment', a free grant of one-quarter of the normal regional allocation, was an option accepted by many serfs and encouraged by the landowners, particularly in the fertile regions where land was most valuable, since it enabled them to retain most of their estate. Not only did peasants who bought their land have to undertake redemption payments for 49 years against loans advanced by the state, but the repayments, with interest at 6 per cent., were too high for their earning power,[2] the valuations of the holdings being excessive.[3] The peasant, moreover, did not, except in some areas, become a landowner, for the title of the land was vested in the mir.

In all there were about 47 million serfs (out of a total population of some 74 million) in Russia at the time of the emancipation. There were two main classes, those belonging to the landlords, to whom this edict of 1861 applied, and those belonging to the state and the imperial family who were emancipated in 1866. Landlords' male serfs, or 'revision souls', and their families, numbered about 22 million people, inhabiting the lands of fewer than 107,000 gentry. Eighty-one per cent. of the serfs were owned by 22 per cent. of the landowners, these being those who owned more than 100 souls. Less than this number of serfs was generally considered inadequate to maintain a family of the gentry.[4] Thus, most of the gentry were, in effect, landowners of very moderate

[1] Florinsky (1953), vol. 2, 891.
[2] Pavlovsky (1930), 77–78.
[3] Robinson (1932), 88; Seton-Watson (1952), 43–44.
[4] Emmons (1968), 4.

substance, while the great landowners spent little of their time on their country estates. In neither case was there an adequate basis for modernization of agriculture, yet a very large potential supply of labour was virtually untapped.

Most of the implementation of the emancipation took place in the decade following the edict, two-thirds of the landlords' serfs having accepted redemption by 1870. Liberal reformists pressed for obligatory redemption to speed up the process, but this was resisted by Alexander II and was not authorized until 1881, after his death.

THE EFFECTS OF EMANCIPATION

The peasants clearly gained little from emancipation. In theory they were free, and in theory they gained some local administrative powers through participation in the commune. In fact, however, the commune was itself the agent of the government, collecting the redemption payments and taxes and administering the hated passport system that had helped to tie the serf to the land since its introduction by Peter the Great and which was not abolished in the emancipation measures. Even the right to work one plot of land continuously, which conferred some feeling of ownership and permitted the peasant to leave the plot to his heir, was limited to a minority of the communes. Elsewhere the system was repartitional, the land being still kept in the old three-field system and divided into strips, periodically re-allocated among the peasants. The average size of all peasant holdings was about 16 hectares but that of the former private serfs only about 10–12 hectares.[1] This was insufficient for a family in most areas and on the more fertile soils holdings were generally smaller. Inadequacy of their land holdings obliged the peasants to work for the landlords, but earnings remained pitifully small and once debts were incurred it was almost impossible for a peasant to repay them.

Many peasants sought the solution to their problems in industrial employment. Those living near the growing factory towns could continue to live on the holding and travel daily to work, but, with the limited transport of the time, this must have been possible only for very few. For most, factory work would

[1] For details see Pavlovsky (1930), 76; Timoshenko (1932), 50; Gsovski (1949), 667–669.

have meant the absence of the man or men of the farm for most of the time and the relegation of the holding to part-time maintenance and the position of an insurance against the time when factory work was no longer possible. In 1900 more than one-half of all Russian factory workers retained a plot of land in their village, indicating the continuing tie between rural and urban life, but gradually a permanent industrial labour force was created and the links were broken.[1] The effects were not entirely harmful because of overpopulation of the agricultural lands, but it was probably the more enterprising who quitted the land in favour of industrial employment.

The indebtedness of the less fortunate and the drift to the towns facilitated the rise of the *kulak* class, the peasants who, through a combination of better management, luck and taking advantage of their less fortunate fellows, bought extra land as it became available and improved their position.

The resentment of the peasantry at the terms of emancipation together with the dissatisfaction the gentry felt at the limitation of their administrative role, ruled out the possibility of a peaceful evolution of the countryside towards modern farming systems. The rioting that immediately followed the Edict of Emancipation declined as the mediators appointed to give effect to the legislation emphasized to the peasants the better effects of the reform and convinced them that this was all that they could expect at that time. These mediators generally sympathized with the peasants and some of them met violent opposition from provincial gentry.[2]

The gentry had their own troubles, notably continued difficulty of financing farm development. Some were seriously disturbed at their removal from positions of responsibility, local administration having fallen increasingly on the communes and on the bureaucrats appointed by the government. There was widespread discussion in assemblies and in journals on the future role of the gentry in society, and on the advantages of abolition of the special privileges they retained in exchange for a voice in government. The gentry had been asked for opinions on the reform of local government, but were given no opportunity to shape the reform. When the principles for the reform

[1] Von Laue (1961), 63–65.
[2] Emmons (1968), 330–331, 334.

were published in 1862, for implementation in 1864, they satisfied neither the liberals, who wanted greater reform, nor the conservatives, who feared the influence that might be wielded by both liberal reformers and the peasantry. The zemstvos, the newly created local councils, however, made outstanding contributions in educational and health services, which were added to their activities in an amendment to the legislation. The work of the zemstvos in agricultural improvement will be referred to later.

For the financing of farm development the gentry had looked to the redemption payments to free them from accumulated debts and to provide capital for development. The redemption payments, however, did not provide as large a source of revenue as had been anticipated by many who had not calculated their real value. State deductions for debts amounted by 1881 to about 303 million rubles out of a total of some 740 million rubles paid for redemption.[1] and depreciation of the currency progressively reduced the value of the income from this source. The landowners now had to pay for their hired labour, and probably few had realized the full value of the enserfed labour they had previously enjoyed.[2] Prices of agricultural products on world markets, especially grain, declined as competition from new lands was felt in Europe. Within Russia, it was the farms of the southern steppes and Siberia, which had been developed with little or no serf labour, that were best able to compete in the new conditions. The gentry in other parts of Russia increasingly resorted to selling off their estates, so that by 1914 they probably held less than half the land with which they had been left after emancipation, and indebtedness had again increased.[3]

Although the task that faced the gentry in adapting to the new conditions after the emancipation was difficult, it is hard to believe that they could not have achieved more in the transition to full commercial farming. They were, however, dominated by their past and many felt that their heritage had been stolen from them as assuredly as did the peasants who had found their new holdings vested in the mir. Furthermore, the gentry were

[1] Emmons (1968), 421.
[2] Pavlovsky (1930), 99.
[3] Emmons (1968), 421–422.

not a class brought up to manual labour or any other form of hard work, such as was required to overcome their difficulties, and too many let their land to peasants rather than attempt to farm constructively themselves. Some, however, persisted and large farms were the basis of the modest improvement that took place in the quality of farming, and provided most of the crops for export.[1]

THE REVOLUTIONARY ELEMENT

Disillusionment with the emancipation was widely taken as proof that only violent revolution could bring thorough and lasting reform. During the 'seventies, the rural areas were penetrated by the new reformers, armed with varying philosophies of resistance and violence. The ideas of Herzen and Chernyshevskiy, who thought that the mir could provide a road to agrarian socialism, guided the young revolutionaries who decided to 'go to the people' in the countryside. They sought to encourage the peasants to unite and seize the landowners' estates and add them to the commune to form a new Slav form of agrarian civilization. They failed to convince the peasants of their own powers, and Plekhanov rejected these ideas and preached the universal redistribution of land. His 'Black Redistribution' group combined with followers of Lavrov, who stressed the need to educate peasant leaders, and from it emerged the Russian Social Democratic Labour Party in 1898. It was to this party that Lenin, who had studied peasant attitudes during his first exile, brought the gospel of uniting the peasants with factory workers, soldiers, sailors and the intelligentsia in Marxist groups to form a sufficient force for the seizure of power. In 1903 the followers of Lenin achieved a temporary majority on the party's executive and the term 'Bolshevik' was born to contrast his group with the Menshevik, or lesser faction. The appropriation of land for redistribution by the peasants was adopted by the R.S.D.L.P. and the Bolsheviks also discussed proposals for collectivization of land and farms.

For the great majority of peasants, however, the old grievances smouldered on without the active stimulation of the

[1] Pavlovsky (1930), 220–221.

revolutionary parties. Outbreaks of violence were sporadic and largely spontaneous, but, in addition to active attacks on landowners, peasants who had contracted to work on estates to supplement their meagre incomes found that they had a new weapon in the strike.

By 1905 the weakness and disorganization of Russia had been made apparent on the international scene by the defeat of the Russian Empire by the emergent Japanese, and returning peasants, after witnessing the humiliation of their leaders, encouraged their fellows at home to press more actively for their rights, however poorly these may have been formulated. In the first and second Dumas, peasant deputies spoke of the continued sufferings of the peasants and repeated the familiar accusation that the gentry had stolen the land. The Peasants' Union was formed in 1905, and in the communes the assemblies were often turned into rebellious political meetings. Nearly 1,500 peasant disturbances were recorded in 1905–6, ranging from thefts and raids to arson and murder. Riots were particularly common in guberniyas with a high proportion of small peasant plots, a high correlation having been found particularly between numbers of riots and the proportion of plots between five and ten dessiatins in size.[1] In the South-western, Ukraine and Central Agricultural regions the average sizes of peasant holdings were only 5·5, 6·1 and 7·8 dessiatins respectively.[2]

The riots between 1905 and 1908 were met by repression throughout Russia. Apart from those killed in the attempts at revolution in Moscow and elsewhere, probably between 3,500 and 4,500 persons were executed.[3] These riots were the culmination of years of unrest in the countryside and by 1905 the government could see that a further major reform would have to be undertaken, but action came only after the risings in the cities as well as in the villages, which added up to the Revolution of 1905.

Following the October Manifesto, containing the government's concessions, elections were held for deputies to the first Duma and peasants were well represented, so that agrarian reform was a major issue in debates. The Tsar and his ministers

[1] Cox and Demko (1967): 1 *desyatina* = 1·125 hectares or 2·7 acres approx.
[2] Pavlovsky (1930), 89.
[3] Sumner (1947), 1961 ed. 115, after Robinson (1932) and others.

were, however, alarmed to find that the peasantry, which they had assumed to be a solidly conservative element, was in fact pressing for power as hard as the deputies from the industrial areas. This Duma, was, therefore, dissolved under the powers retained by the Tsar, but the second Duma proved no less radical. The election system was then altered so that the representation from the workers and peasants was sharply reduced, and the third Duma (1907) was conservative enough to survive for its full term. Meanwhile, the Prime Minister, Stolypin, was working on a system of agrarian reform which, it was hoped, would effectively counter the unrest, and this was introduced in 1906.

Thus, another reform was imposed from above and did not immediately produce peace in the countryside. Nevertheless, the measures introduced did provide the basis for eventual transformation of Russian agriculture from the still basically medieval system, which had survived emancipation, into a modern system of compact, privately owned farms. Stolypin was emphatically opposed to the commune; the private small farm was his formula for stability. The establishment of such holdings was to be achieved by removing most of the commune's powers, including those in local taxation, and ending the redistribution of strips of land. The eventual aim was the creation of consolidated holdings but, as this could not be achieved immediately throughout the country, there was to be an intermediate stage which made the strips hereditary, if they were not already so. When a holding was consolidated in one enclosure (*otrub*) the family could build a new farmhouse on it and move away from the large nucleated village. The dispersed farmstead (*khutor*) thus became a common feature of the landscape.

To assist the emergence of a class of more prosperous peasants finance for development was made easier to obtain. A State Peasants' Land Bank had been created in 1883 but its impact was negligible in its first decade. From 1895 to 1905 it was more effective and in 1905 it was authorized to purchase land on its own account for peasant settlement and interest rates were reduced.[1] migration was made easier and the resulting availability of holdings enabled the more prosperous peasants

[1] Pavlovsky (1930), 154.

to buy up land. It was also made easier to rent land and these conditions strengthened the position of the more prosperous peasant, or kulak. The holdings of the kulaks compared favourably with those of the more progressive landlords and often showed the beneficial effects of fertilizers, a modest degree of mechanization and improved seeds and livestock. These two classes farmed increasingly for the market. The land held by the nobility and gentry was, however, in many cases deteriorating. The problems created by absentee landlords, living in the cities on the proceeds of their land, had become acute. Some of the land had been sold to the more progressive farmers but much remained to be farmed by poor peasants, cheaply hired labourers and tenants who were denied the capital required for improvements.

Those peasants who were confined to the tiny holdings received under the Emancipation Act were relatively worse off. Although the redemption payments were cancelled in 1905, the repayments during the previous fifty years had been a serious drain on their minute incomes. In addition, they had suffered the full burden of taxes on consumer goods such as sugar, tea, tobacco, cotton and iron goods, as well as on the vodka that was a curse of the countryside. Generations of peasants had mortgaged their crops while they were still growing for the supplies of vodka their landlords sold them at exorbitant prices, and alcoholism was an old and acute problem.

In the half-century following emancipation changes had taken place in the priority given to different items of output. Production of wheat had greatly increased while rye was relatively less important. New wheat lands east of the Volga accounted for much of the change, but rye was giving way to industrial crops and wheat in the central chernozem area. Wheat was an important export crop and the demand for sugar beet for the factories, established in the nineteenth century, had led to a marked reorientation of production in the southern regions. Other crops which had been developed in quantity for the industrial or export markets were cotton, sunflower, tobacco and potatoes, but flax and hemp were declining.[1]

[1] Parker (1968) details many of these changes, as well as summarizing the changes of earlier periods, based on Lyashchenko (1947–56), Florinsky (1953), Baranskiy (1950), Robinson (1932) and the writings of many contemporary observers.

REGIONAL DEVELOPMENT UP TO 1917

Greatly contrasting fortunes had affected the various regions between the Napoleonic Wars and the Revolution of 1917. In some areas, new ideas in the science and practice of agriculture had penetrated into Russia, as had the reforming creeds of democrats and revolutionaries in politics, and the growing commerce and industry of the towns had stimulated commercial agriculture. New lands were developed in Siberia, on the Black Sea steppes and the more recently conquered Caucasian and Central Asian territories. Elsewhere, however, conditions had deteriorated even further under the extortion of absentee landlords, the dead hand of the commune and the abject poverty of the peasants.

The North European Area

Early in the nineteenth century there were areas around Arkhangelsk, Kholmogor and elsewhere north and east of St. Petersburg where agricultural improvements from the period of Catherine II had resulted in substantial changes. The crossing of Kholmogor cows with European bulls had led to the development of cattle with higher milk yields and better beef qualities than in most parts. More generally, however, shifting agriculture remained. In spite of the labour expended on constant clearance of regrowth and the difficulty of weeding, forest clearings were relied upon for poor crops of rye, barley, and oats. On the southern margins of this zone, flax was an important cash crop throughout the century. Potatoes became important, to some extent competing with flax, since they could be used for human subsistence and animal fodder or sold for cash, as the season dictated.[1] The areas were not attractive to the nobility and gentry and the proportion of serfs to free peasants was relatively low.[2]

The Baltic Provinces & adjacent Northwestern Areas

The accessibility of these areas from the west facilitated the early adoption of many improvements. German landlords in

[1] For a more detailed regional treatment of the agriculture of European Russia at this time, including land use statistics and land prices, see Pavlovsky (1930), 38–59 and maps.
[2] See map, Parker (1968), 234, based on Lyashchenko (1939) 1949, 311.

Livonia reorganized their estates early in the nineteenth century and serfdom was abolished earlier than elsewhere in the empire. Improvement of pasture made possible better exploitation of the cool, damp climate. Cattle, imported from the Ukraine, were fattened for sale in St. Petersburg, good sheep were reared on the coastal salt marshes for semi-fine wool, and flax was a reliable cash crop. Rye and other grains were produced for export and for the manufacture of vodka as well as for local consumption. Riga became famous for its beer, which further improved the market for grain. Crop rotations were fairly general, the three-field system having been widely abolished and the cereal yield was double that of neighbouring Russian provinces.[1]

Belorussia

Conditions deteriorated in southern Lithuania and were extremely bad in the marshy and morainic areas of Belorussia. Drainage was urgently needed but, after some progress in the 1870s and 1880s, improvement work slowed down.[2]

In the better parts of Belorussia and in the adjacent drier country of western Russia, for example, the Smolensk area, substantial progress was made before the end of the century, and Stolypin's reforms effected further improvement. Sown pastures, with clover, were introduced, and cattle numbers and the cultivated area increased considerably. Industrial development and railway communications provided a high degree of stimulus.

The Central Podzol Region

The zone of mixed forests around Moscow, extending northwards and westwards to the previously described areas, southwards to the forest-steppe and eastwards to the Urals, was poorly developed agriculturally even at the end of the nineteenth century. During the century, shifting, forest-clearing cultivation gradually became less common as population pressure increased. The three-field system was the normal method of exploitation, with rye, oats and buckwheat the most important

[1] Parker (1968), 238.
[2] French (1963), 53–54. See also Ch. 11.

grains, and flax the major industrial crop. Before the emancipation serfs formed a high percentage of the population, but the obrok system prevailed, the money dues extracted from peasants who went to work in the industrial towns or undertook domestic manufacturing tasks being more welcome to the landlords than the full quota of work on the poor soils.[1]

Immediately after the emancipation of the serfs, the development of the railways enabled grain to be brought into the industrial areas from the south, and the latter half of the nineteenth century saw a changeover to the growing of potatoes, vegetables and fodder crops for livestock rearing. Dairying, in particular, developed near the industrial towns, and towards the end of the century sown grasses and clovers were being adopted by the more progressive farmers. High prices for land reflected relatively good returns and the pressure of population.[2] The sown area, however, declined over much of this region between 1860 and 1913, in contrast to its expansion in surrounding regions.[3]

The Forest-Steppe Region

The fortunes of the forest-steppe changed drastically during the nineteenth century. At the beginning of the century, it was emerging from the centuries of wars and raiding that had retarded its economic development. Grain growing was soon expanded, favoured by soils of the chernozem type, sunny climate, relatively mild winters and comparatively easy access to the growing towns of the central industrial region and those within or on the edge of the forest-steppe itself (Kiev, Kharkov, Voronezh, Tula, etc.). In 1810 a traveller wrote, 'You travel for miles and miles and see nothing but corn'.[4] Apart from wheat, the main crops were barley, oats, millet, peas, lentils, flax, hemp, sugar beet and poppies, with buckwheat on sandier soils.[5] Orchards were planted, especially in the southern areas, and livestock rearing became more a farming and less a semi-nomadic pursuit. Merino sheep were introduced into this

[1] See Parker (1968), 234–235, for maps based on Lyashchenko.
[2] Pavlovsky (1930), 44–45.
[3] Parker (1968), 286 (map); Lyashchenko (1939) 1949, 451.
[4] Clarke (1810), vol. 1, 189, quoted by Parker (1968), 160.
[5] Storch (1801), vol. 2, 232, quoted by Parker (1968), 160.

region as into the southern steppes during this period of development.

Sugar beet was introduced into the forest-steppe region early in the century, the first factory to process it opening at Tula in 1802. In the middle decades of the century the beet-sugar industry grew rapidly, mainly in the forest-steppe and northern steppe areas, where both climate and soil suited the crop excellently.[1] Most of the refineries were operated by landlords on their estates and this agricultural-industrial link was, with the grain export industry, the main reason why the landlords in this region were reluctant to let their serfs leave the land in return for obrok payments. The proportion of serfs to the total population was high, and the percentage on barshchina was generally over 60, rising to 90 per cent. in some areas, as in the steppe lands further south. Furthermore, the landlords exacted the maximum labour from their serfs, often more than the customary three days per week. A Tula serf owner who ceased to demand barshchina on Sundays was accused of coddling the peasants.[2]

Along with the exploitation of the serfs went the continuance of the three-field system, with repartition of holdings, over-cropping, reliance on bare fallow to restore fertility, and ignorance or unwillingness to recognize the rapidly developing soil erosion. Banks and gullies too steep for ploughing were grazed much too heavily, leading to destruction of the cover and gulleying as well as wind erosion.

Emancipation of the serfs was resisted by many of the land-owners here because of their labour value but, since the land was valued by them even more, they retained the maximum and urged the peasant to take the 'beggars allotment'—the free plot of one-quarter the size that would have otherwise been granted. Grinding poverty became in this most fertile region more wide-spread than anywhere else in the Russian Empire. Outlets for industrial work within the region were few, so there was much permanent or seasonal migration. The railways proved a very mixed blessing to this region since they opened up the grain trade to the more southerly and newly developed areas leading to a retreat to subsistence cropping by the peasants of the forest-

[1] Parker (1968), 246–247 (map, 247).
[2] Blum (1961), 446.

steppe. There was no room for pasture and hence livestock were few and manure scarce; a cycle of decreasing fertility became general. The wooden plough and harrow remained in use and peasants had little prospect of obtaining newer implements when they were falling more and more into debt. Many even had to sell their horses, so losing the means of providing their own animal haulage. Stolypin's reforms improved conditions for some but increased migration, as the more prosperous and commercially-minded kulaks bought out the poorer peasants.

The Steppes

It was even truer of the steppes than of the forest-steppe that conditions largely precluded agricultural development before 1800. The Don Cossacks and some other communities had built up great herds of cattle, while to the east the Tatars were still semi-nomadic livestock herders. Cattle were fattened in the Ukraine and exported on the hoof to the Baltic region.[1] Sheep were numerous, with the long-tailed Circassian breed predominant, but, as already noted, the Merino was introduced in the first decade of the nineteenth century.

As new landlords perceived new ways of exploiting the steppes, the landscape began to change more rapidly than elsewhere in the Russian Empire. Grain and sugar beet became staple crops. West of the Dnepr, in former Polish territory, serf labour continued until the emancipation, but to the east serfs were fewer and hired labour became more common, though over 90 per cent of the serfs were on barshchina, and were exploited unmercifully. Labour was, however, relatively scarce, even though many peasants migrated seasonally from the forest-steppe to work on the farms. Because of labour shortage and comparative freedom from the traditional ideas that governed the maintenance of the commune and the three-field system, agricultural techniques progressed faster than in the central regions. More machinery was introduced and teams of oxen were used to draw heavy ploughs. Steam engines were introduced for stationary purposes on mills and in farmyards as well as for transport. The railways gave access to the markets in the industrial areas and to the ports for export. Other crops

[1] Parker (1968), 160.

were produced as industrial demand increased, sunflower for oil becoming a major crop between 1900 and 1914.

The transformation of the landscape with the growing of crops for export and industrial use continued into the Crimea and across the Don to the north Caucasus steppes. The extension of the railway from Rostov to Baku opened up a fertile region for development, and colonists came in from Germany and the Balkan countries. Shortage of labour compelled farmers to adopt machinery, and the Kuban established a reputation for high farming which distinguished it from longer settled areas, a distinction which still remains.

Siberia & the Far East

Primitive cultivation was carried out by many tribes inhabiting Siberia before the advent of the Russians, and peasants were an important element in the Russian colonization east of the Urals. Armstrong writes: 'For all the hunters, the exiles and miners about whom one hears, Russian Siberia was primarily an agricultural country.'[1] In the seventeenth century, grain was grown in the plains of the Irtysh around Tyumen, and agriculture was practised almost as far north as latitude 62 degrees in the Yenisey and Lena valleys.[2] Rye, wheat, barley, oats, peas and hemp were grown from an early date, adding to the knowledge of the natives not only new crops but new techniques, including that of ploughing.

Most of the peasants in Siberia were 'state peasants', living in villages in which some fields belonged to the government and some to individuals. They had to work part-time in the government fields but supported themselves from their own plots. Later, work in the government fields was replaced by obrok payments. Thus, it is not true that there was no serfdom in Siberia, but it operated in a modified form, with no buying and selling of serfs.[3]

Zones in which agriculture was practised show gradation from north to south, as in Europe and Russia, but with only sporadic development in the north and east because of the inhospitable conditions and distance from major settlements.

[1] Armstrong (1965), 72.
[2] Armstrong (1965), 73–78.
[3] Armstrong (1965), 75.

Tatar domination kept Russian agriculture from the steppes in the eighteenth century, while the Chinese kept them from the Amur valley until the 1860s. A little agriculture was carried on in Kamchatka from the eighteenth century, but development in the Far East was negligible until the left bank of the Amur was secured and Vladivostok founded. The Trans-Siberian Railway was a major factor in encouraging cultivation and stock rearing early in the twentieth century. In particular, it helped the dairy industry, which had been established much earlier in the forest-steppe in western Siberia, and it provided new markets along its whole route. Many co-operatives were formed to produce and market butter, which was exported in large quantities. The peasants generally owned more livestock than did those in European Russia and the average holding was nine times larger.[1]

Transcaucasia & the Caucasus Ranges

While the development of the north Caucasus lands followed that of the Don steppes and was primarily an extension of grain growing and livestock rearing, that of the Transcaucasian valleys was quite different. In the first place, Russian conquest or incorporation did not take place until after 1800, and secondly, the winters were much milder. Vines, mulberries, olives, figs, nuts and various orchard fruits as well as wheat, melons, water melons and other crops of Mediterranean or sub-tropical origins were grown there before the Russians took over the princedoms of the Kura and Kolkhid lowlands. The new overlords and migrants, who included Volga Germans as well as Russians and Cossacks, developed wheat, cattle and sheep farming as they did elsewhere, but they also intensified the production of more specialized produce, including grapes, tobacco and citrus fruits. Cotton was introduced and its production expanded rapidly before the end of the century, the growing of sunflower seed was developed to meet demands from other areas, and tea was introduced before the First World War. The mountain areas, occupied militarily during the century were less attractive and were largely left to the native peoples who had retreated to them.

[1] Treadgold (1957), 207–212.

Livestock rearing there was carried on partly, at least, on a transhumant or semi-nomadic plan.

In spite of the variety and richness of the produce, the average peasant existed on a meagre wage or a tiny holding of land which offered him only a bare subsistence, while the landlords, whether Russian or assimilated Georgian, grew wealthy on the proceeds.

The Crimea

An agricultural pattern somewhat similar to that of the Caucasus developed in the Crimea, with wheat the main crop on the steppes in the north and orchards, vineyards and vegetable gardens in the warmer, moister south.

Central Asia

The final region, or group of regions, to be briefly noted in this historical section, is that east of the Caspian Sea and south of Kazakhstan. These were the last major regions added to the Russian Empire, but their conquest in the late nineteenth century brought under the Tsars peoples descended from civilizations which were highly developed before Kievan Rus emerged as an embryo Russian state. When the Russians conquered them, however, these ancient states had fallen into decay, having warred among themselves for centuries. Furthermore, the Moslem faith had kindled in the hierarchy no kinder an attitude to the poor than was evident in the Russian lands.

Before the Russians came, the oases of Central Asia supported agriculture based on fruits and grains, and the colonists showed the natives that cotton could be more profitable than silk, or, better still, that cotton could be grown between rows of mulberry bushes that would shelter it. Development was hindered by smallness of holdings and the lack of education and commercial spirit among the natives, and the Russians found it hard to recruit wage labour. The irrigation network however, was steadily expanded, and, in 1900, Tashkent became connected by rail to Orenburg and the main Russian railway system. Marketing was thenceforward simplified and production increased rapidly.

The Kazakh nomads who had grazed the steppes bordering

Siberia, however, lost their lands to the Russian farmer, who extended wheat and barley crops and the raising of sheep and cattle southward to the limits imposed on sedentary agriculture by the desert. The Kazakhs retreated into the desert areas, retaining their nomadic way of life until after the Revolution. They were among the last to yield to collectivization.

* * *

By 1914, improvement, aided by these reforms and increased availability of better seeds and livestock, fertilizers and machinery, had spread with much increased speed throughout the country, but especially in the Ukraine, north Caucasus and Siberia. Since 1906, 1,048,007 individual farms had been established but this represented only about 8 per cent. of the total number of households.[1] The majority were still worked in three-field systems dominated by the communes, which had resisted complete reform, even if the strips of land were no longer redistributed. The nobility and gentry, despite increased sales of land to peasants, still held an average of twenty-five times as much land per family as the peasants. The plight of the poorer peasants, especially in the congested areas of the central chernozem and middle Volga provinces, remained pitiful.

The reforms promoted by Stolypin might eventually have achieved their object, but the entry of Russia into the First World War destroyed this hope. The loss of men to the armies and the breakdown of supplies required to maintain production put intolerable burdens on the women and the very young and very old who remained on the land. Arrangements for marketing produce also deteriorated as the war went on, and hunger in St. Petersburg, Moscow and other large cities contributed materially to increasing opposition to the Tsar and his conduct of the war.

The February Revolution was achieved by liberals whose main preoccupation was to maintain order and continue prosecution of the war. They assumed that the removal of the Tsar and the institution of more democratic government would persuade the people to continue to fight alongside

[1] Gsovski (1948–49), vol. 1, 683–684.

the Allies. Little could be done in the disturbed circumstances of the following months to speed reforms in the countryside, but the new government undoubtedly weakened its position by delaying pronouncements on its intentions while it awaited the holding of a constituent assembly to confirm its position. It thereby unwittingly committed the country's agriculture to the experiment of collectivization in the service of a Soviet state.

THE PRELUDE TO COLLECTIVIZATION

The first concern of the Bolsheviks, on seizing power, was the reorganization of production, trade and distribution on socialist lines. A decree confiscating the lands of the church, monasteries and landlords generally was among the first of Lenin's acts. The doctrine of socialization of land was effected in the decree of February 19, 1918, which abolished all private ownership of land.[1] Following Marxist and Bolshevik theory, the formation of large farms, both of collective and state-organized types, was undertaken on the limited scale practicable in the confused conditions of the aftermath of the Revolution, but there was little to prevent the peasants seizing what land they could.

In July, 1918, the Congress of Soviets adopted a resolution calling for reorganization of communes into larger farms and the development of state farms. Shortly afterwards, the Commissariat of Agriculture issued a decree to initiate moves in this direction, but the law promulgated in February 1919, to give effect to the proposed scheme of reconstruction, remained ineffective. The peasants were bitterly opposed to surrendering land to collectives, and as the civil war flared up and spread across the land systematic reorganization became impracticable. Many risings against the Bolsheviks occurred in the countryside whether or not consciously in support of the counter-revolutionaries. During this period of War Communism (1918–21), peasants were required to make compulsory deliveries of food to the state but generally resisted doing so, and the supply position continued to deteriorate and famine conditions again gripped the towns. In a retreat from principle to restore production and supply,

[1] Gsovski (1948–49), vol. 1, 691–692.

the New Economic Policy (1921–28) made concessions to private enterprise. Taxes in kind or money replaced requisitions of agricultural produce. The more successful peasants found that they could rent land to extend their holdings and hire labour and increase private livestock and equipment, although the state remained the only official landowner and the Land Code of 1922 expressly forbade purchase, sale, mortgage or bequest of land, which was generally resented by the peasantry. In spite of the reorganization and stabilization of the currency in 1923–24 peasants were reluctant to part with their produce. Evasions of tax payments mounted and speculation in grain and other commodities became rife. The behaviour of the peasants strengthened the hands of those advocating total change from the landscape of small, ill-equipped, private farms. It was argued that no modern agriculture capable of serving the developing Soviet state could be based on such a system, or lack of system, and that the only solution acceptable to communism lay in collective or state farms of large size, corresponding to industrial evolution in the towns.

THE COLLECTIVIZATION DRIVE

Under War Communism and the N.E.P. the appeals of the Bolsheviks to form collective farms met with little response from the peasants. The Land Code of 1922 offered the village communes a choice between adherence to pre-revolutionary forms of land tenure and adoption of co-operative, collective or communal systems.[1] One of these forms was the *artel'* (a name used for a traditional form of workers' co-operative), which later became the officially adopted model. In general, only idealists and the poorest peasants were interested. Some of these, however, went so far as to divide virtually all property communally and to set up communal dining halls for all members and dwelling quarters for bachelors and orphans. Religious sects founded some of the communes but these did not survive the main wave of collectivization and became incorporated with others with no religious background.

The Society for the Joint Cultivation of Land (TOZ)[2] also promoted collectivization on a voluntary basis. In farms

[1] Gsovski (1948–49), vol. 1, 698ff.

[2] *Tovarishchestvo po sovmestnoy obrabotke zemli.*

of this character, the value of implements, livestock and other productive resources could count towards the share in the proceeds received by the individual member, which was a source of dissatisfaction to the poor peasants, who had little but their labour to contribute. Members were free to withdraw from such collectives, which meant that the farms were liable to find their productive capacity impaired at any time. As long as N.E.P. lasted, there was little incentive for a peasant to work in a collective. While the Bolshevik leaders believed in collectivization, most of them, following Lenin, considered that it could be achieved only slowly and with the peasant's co-operation. In spite of the difficulties of extracting produce from the peasant there was little expectation, even after the death of Lenin, of a forced collectivization.[1]

Stalin, however, blamed the continuing grain crisis on the capitalist peasants, the anti-communist kulaks who had been encouraged under N.E.P., and decided to eliminate them. In 1928 he made a declaration to the Central Committee, justifying a policy of force by the need to reorganize farming to provide the means to industrialize the country. The attack on the resisting peasants grew during 1929 with the backing of the collectivization scheme adopted as part of the Five Year Plan.[2] This originally required collectivization of 20 per cent. of the sown area of the country during its operation from 1928 to 1933. In 1930, however, he ordered greater haste to complete collectivization of the grain areas of the Volga and north Caucasus in 1930 or early 1931 and the total liquidation of the kulaks began.[3]

The poorer peasants were encouraged by the local Communist and Komsomol cells to band together in Committees of the Poor which could provide organizations to attack the kulaks. Property was requisitioned and distributed among the poor or politically active peasants. Some of the buildings of the kulaks who resisted and were killed or exiled formed the nucleus of the communal buildings of the early collective

[1] Dumont (1964), 34.

[2] Interpretation differs on the justification for, and the timing of, decisions on forced collectivization. See, for example, Lewin (1965), Narkiewicz (1966), Karcz (1967) and Narkiewicz (1969).

[3] Conquest (1968) summarizes the events of this and the succeeding phases of collectivization from Danilov (1963) and other authorities.

farms. Villages resisting collectivization were visited by Communist Party officials, who tried persuasion and, if this failed, called in troops and the poorer peasants to requisition land and buildings and create an artel. Differential taxes levied on those who continued to resist finally drove almost all households to join the collectives.

In 1928, collective farms produced only 3·3 per cent. of the Union's agricultural output, according to official calculations.[1] In fact, only in the production of sugar beet, of which they were responsible for 33·4 per cent. in 1928, had they any significance. They produced 3·5 per cent. of the country's cotton, 2·7 per cent. of the grain but 0·6 per cent or less of the milk, meat and eggs.[2] By the end of the Five Year Plan, about 60 per cent. of holdings had been collectivized, three times as many as envisaged for the plan at the beginning of the campaign. The excesses committed in driving the peasants to collectivization at this rate caused widespread dislocation of crop production and destruction of livestock by the unwilling peasants. Later it was revealed that between 1928 and 1933 livestock numbers fell by some 27 million cattle, nearly 64 million sheep and goats, 18 million pigs and 15 million horses—losses amounting to more than 40 per cent. of the 1928 national herd of cattle and horses and two-thirds of the sheep, goats and pigs.

By this time the average collective farm consisted of about 70 peasant holdings, there being 210,600 collective farms, embracing 14·7 million households, in 1932[3] The members were expected to work 100–150 days on the co-operative enterprise and the rest of the time they were free to work on the personal plots of land which they were allowed. From 1930, payment by the labour-day unit (*trudoden'*) was standardized, replacing the varied systems used on the earlier collectives and communes.[4] Machine-Tractor Stations had been established to meet the needs of the farms for cultivation, almost all of which had hitherto been met by horses and draught cattle. There were 2,446 such establishments at the end of 1932. In addition there were 4,337 state farms.

[1] *N.kh. SSSR 1968*, 48.
[2] *Strana Sovetov za 50 let*, 121.
[3] *N.kh. SSSR* (1956), 100.
[4] See Wronski (1957), 14–25.

The first state farms were created immediately after the Revolution, using the more valuable assets of some of the private estates, such as purebred livestock, to form a basis for research and development They were also used to continue the large scale production of sugar beet in which large estates had specialized. State farms were out of favour in the N.E.P. period, but the grain crisis led to the creation in 1928 of large mechanized farms to bring into cultivation some of the unused lands of eastern and southern areas—a forerunner of the virgin land scheme of the 'fifties, which encountered similar difficulties in semi-arid environments. State farms were formed also to promote livestock farming and other specialized branches of agriculture.[1] The state farms were supplied with machinery on a comparatively generous scale but were widely criticized for wasteful use of resources and general inefficiency, particularly having regard to the poor resources available to the collective farms.

A description of a collective farm at this time indicates its extemporary nature:

> Our kolkhoz was built on the site of the former commune, . . . the farm and administrative buildings . . . were constructed from sheds which had formerly belonged to kulaks, the farm buildings of collectivized peasants, and other miscellaneous sources. Tombstones and stone crosses from the cemetery were used for the foundations of the buildings; for the roofs, the sheet-iron was ripped off the former kulak dwellings on the kolkhoz land. Willow and linden wood, of which the village had an ample supply, was also used in the farm's construction. By the summer of 1931, the kolkhoz had its own stud farm with space for 120 horses, a large barn for grain and a steam-operated flour mill.[2]

Only simple implements were owned by the collectives, the MTS being responsible for providing all services requiring heavy and complicated machinery. The MTS were widely regarded as centres for the Communist Party's control of the countryside, but there was sound reasoning in their establishment to develop mechanization of farming. Few collectives

[1] Volin (1951), 69–80 described the development of state farms with a selection of data from pre-war Soviet statistical handbooks and other authorities.

[2] Belov (1956), 11.

could have afforded to purchase a wide range of machines, even had they desired. In fact, there was widespread opposition to the use of tractors among those who had been brought up with horses and working cattle. The value of the manure, especially in the absence of artificial fertilizers, the adaptability of draught animals which also provided food and skin products, and the fact that they could be maintained on a farm's own produce and not be dependent on deliveries of oil and spare parts all commended them to the collectives. Mechanization, however, had to be introduced as a first step towards modernization of the farming system and the freeing of grain production for human consumption, and this the MTS could promote.

The MTS failed, however, to gain the confidence of farm committees because of inability always to provide services when required and to carry them out properly. Furthermore, farms had to pay highly for the services of the MTS from their output. They could endeavour to carry out the work with their own animals and implements, but to discourage this, punitive levies were made on their produce. As chosen instruments the MTS continued to grow and by 1940 they numbered over 7,000, and controlled 435,000 tractors, the latter figure being nearly six times as many as in 1932.

The collectives also grew in number and in size and by the end of 1940 they numbered 235,500, averaging 81 households. State farms had decreased slightly in numbers to 4,159 but their share of the sown area had increased to 8·8 per cent. of the total. The collectives held 78·3 per cent., personal plots of all kinds added 3·0 per cent. inside the collectives and 0·5 per cent. outside and 9·4 per cent. remained in the hands of individual farms not as yet collectivized.[1] Thus, the 25 million or so peasant holdings of 1928, averaging 15 hectares in size, had been largely consolidated into large collectives averaging about 1,600 hectares of land, of which about one-third was arable.

By this time the collectives had been able to demonstrate that they had some advantages compared with the tiny, undercapitalized holdings they had replaced. Ploughing and sowing could be completed more quickly with machinery,

[1] *N.kh. SSSR* (1956), 110.

which was especially important in semi-arid areas, irrigation facilities could be developed and used more effectively and the benefits of science could be brought more easily to cropping and livestock rearing. After the famines of the years that followed the First World War and the destruction caused by collectivization, most areas had increased their productivity and the average farm worker was probably living better than before collectivization. Even a disillusioned and embittered exile, who catalogued the sufferings of the collective of which he had been chairman, wrote that, in general, from the mid-1930s until 1941, the majority of kolkhoz members in his area (in the Ukraine) lived fairly well and were never in need of food or clothes.[1]

On the eve of the Nazi invasion of the Soviet Union, the sown area had been increased to 150 million hectares, compared with 113 million in 1928 and 118 million hectares (within the contemporary frontiers) in 1913. Livestock numbers had not recovered to pre-collectivization levels but this achievement seemed within sight, with higher productivity in milk, meat and wool from the improved stock, when war again struck. Large areas were quickly lost to the invaders, a 'scorched earth' policy was applied and livestock and machinery driven away to the east and centre of the Soviet Union. The drive to sustain production resulted in the return to peace being made with total numbers of cattle and sheep only some 13 per cent. less than before the war, but the number of pigs was more than halved, and the productive resources of the farms were very low.[2] The loss of life in the Soviet Union during the war had amounted to some 20 million people, and so labour was short and an immense amount of the work of recovery fell on women and on the elderly.

At the end of the war many peasants hoped for the restoration of private farms. Indeed, in some German-occupied areas this had been carried out, but had again been reversed by the invaders who had accompanied the return to collectivization by brutal repression. Stalin soon disabused the peasants of any hope of private farming and began redeveloping the

[1] Belov (1956), 18.

[2] For an account of Soviet agriculture of the war years with many statistical tables, see Arutyunyan (1963).

collective system to meet the great disparities between agricultural production and the needs of the state.

As in the early days of collectivization, much improvisation was necessary during the years of reconstruction. State taxes in the form of compulsory deliveries of produce were heavy and the resources of farms could be built up only slowly. The supply of machinery was slow because most factory capacity was devoted to even more urgent forms of production. Farms had to trade on the black market for some of their needs, often carrying out direct exchanges of crops, livestock produce or even live animals for materials and spare parts which the proper channels failed to supply. Even theft was condoned by farm managements:

> In many cases implements had to be manufactured at home on a makeshift basis. Harrow teeth were made from railroad spikes; sometimes the kolkhoz bought these illegally from the railroad gangs but more often they were stolen. Steel for sickles and scythes was taken from wrecked tanks.[1]

Slowly, the position improved. In particular, mechanization was stepped up. In 1950 there were 595,000 tractors in use it was reported, compared with 531,000 in 1940, and 211,000 grain combines, compared with 182,000 on the eve of the war. About 60 per cent. of all harvesting was said to be done by machine and almost all ploughs were mechanically operated. This position, however, was far short of the level of mechanization in many western countries and agriculture in the Soviet Union in the early 'fifties was neither efficient nor well equipped by western standards. In the absence of adequate productive facilities, much attention was given to remedies through political manoeuvres and repeated reorganizations, accompanied by technical advices which were often only panaceas.

[1] Belov (1956), 25.

CHAPTER 3

Economic and Political Factors

The effects of these two factors, or, rather, groups of factors, cannot be separated. In any economy, the producer seeks to maximize his profits and this requires production of the goods most in demand within the range for which environmental conditions are suitable. If exchange with other regions is adequately organized and equitably conducted, there is also gain for consumers in such adaptation to environment. Such an ideal state is modified by social pressures such as the stratification of the society, levels and distributions of incomes and resources, attitudes to work in general and to particular occupations. Political factors modify responses more directly, and measures such as tariffs and import quotas alter the individual's most remunerative production pattern. Governments must seek to ensure the security of their subjects and to do this they may stimulate certain types of production and maintain forms of trade with other countries to produce a pattern of production which may be far from the most economic.

In the Soviet Union, political considerations are prominent in shaping the economy, and socio-political ideals are deeply imbedded in the forms of productive enterprises. In agriculture, these are most obvious in the collectivized form given to farming organization. They loom large also through state procurement of output as well as state distribution of many inputs and the state-operated transport service.

Decisions on such economic matters as investment and procurement prices reflect the varying successes and failures of the different political groups, for despite the apparently united front presented by the Soviet government to the outside world, there is continuous manoeuvring and in-fighting among rival political interests. This is well illustrated by the changes in stress given to the conversion of collective farms into state farms, the amalgamation of collectives and the closure of the Machine-Tractor Stations (MTS). Although the last move reduced the indirect control over farming by the central government and Communist Party, the amalgamation of collectives facilitated the formation of Party cells in the rural areas, while the extension of the sovkhoz form increased direct ministerial control over farming. The changes wrought in the structure of Soviet farming through the relative importance of state and collective farms are of prime importance, embodying many of the economic, social and political changes in Soviet attitudes towards farming, and these changes will be examined in this chapter. First, however, the direct action of the Soviet administration on the spatial pattern of production must be noted.

Coincident with the reconstruction of the organizational basis of Soviet farming through collectivization, Yakolev, People's Commissar for Agriculture, told the 16th Party Congress in 1930 that the spatial pattern of agricultural production must be changed to permit a much greater degree of specialization. Land which was insufficiently used was to be developed according to its natural potential, with low-yielding crops being replaced by those more suited to the regional environment. In the cooler areas, marginal for grain production, there was to be more emphasis on fodder crops and livestock husbandry, flax and potatoes; in the forest-steppe, more specialization in sugar beet; in the steppes, more grain and in Central Asia, more cotton.[1] Such plans were necessary to exploit better the natural conditions to which the fragmented peasant farming had been ill-adapted because of the tendency to produce a wide range of products for local self-sufficiency, often on highly unsuitable soils. The large state and collective farms, subject to direction throughout the land, could be made to specialize. At the same time there were limits to the degree of specialization that would be economic, having regard to the amount of interchange of products required, the capacity of the transport system and the need to recognize the dangers for regions, and for individual farms, of excessive specialization.

In the event, some retreat from the objective of specialization

[1] Jackson (1961).

was needed in the chaotic conditions which resulted from the collectivization drive and increasing transport problems. The advantages of greater regional self-sufficiency were admitted and the second Five Year Plan (1933–37) promoted a northern wheat area to reduce the dependence of the central and northern industrial cities on southern grain.[1]

In the same period there was increasing concern over the soil exhaustion caused by excessive cropping in grain areas. Mineral fertilizers were in extremely limited supply and there were insufficient livestock to provide adequate manure. In these circumstances Stalin was won over by scientists who favoured the *travopol'ye* approach to land use, and the changes which were then forced on farms throughout the Soviet Union illustrate the power of Stalin and his Party supporters to dictate technical practices.

The travopolye system relied on grass leys to build up fertility through improving soil structure, which, it was argued, minimized the need for application of fertilizers or other ameliorative measures. Its original advocate, Vasily Robertovich Vilyams, the son of an American railway engineer, developed his theory late in the nineteenth century, when agriculture in Russia was extremely backward, and rotation of crops was almost unknown. Observing the rapid development of farming productivity in Denmark and other western European countries, he fastened on the grassland stages in rotations as the reason for the rapid improvement. He argued that a good crumb structure, permitting aeration of the soil and maximum root development, was essential and attainable only through long courses of perennial grasses.[2] The importance of structure is, of course, widely accepted, but Vilyams differed from most soil scientists of the time in neglecting the importance of fertilizers and failing to recognize other means of raising soil fertility, such as the use of legumes in rotations.

Russian soil scientists were then in the forefront of developments in pedology and soil geography, but Vilyams argued that their work did not help the peasant, who had neither the knowledge nor the resources to convert their findings on the soil profile and its chemistry to practical cultivation

[1] Jackson (1959).
[2] Vilyams (1948–53).

in the fields. In fact, Russian agriculture was tied to primitive methods. There was still widespread use of shifting cultivation incorporating re-use of land after long fallow (*perelog*, as opposed to *zalezh'*, where there was no planned return to the same plot.)[1] Two-field and three-field systems, in which fallow allowed the soil to recover after grain cultivation, were still normal even on relatively well organized estates. Only the most advanced farms used a rotational (*plodosmennaya*) system depending on root and green crops to restore fertility. Extension of such systems was supported by many soil scientists before as well as after the Revolution, but the stress that Vilyams put on the merits of grass in the rotations brought him into conflict with the majority, even though they agreed on the merits of planned rotations.

After the Revolution, Vilyams advocated the adoption of travopolye throughout the country. At the All-Union Conference on Drought Control in 1931 approval was given to a trial of the system by selected state farms and MTS. Various other cropping plans were tried during the early years of collectivization, particularly as emphasis was increasingly placed on the virtues of greater regional self-sufficiency. In 1933 all regions were ordered to produce their own meat and potatoes and to plan rotations but, by 1937, only one-third of all farms had started using rotations and one-third had not even put them on paper.[2]

Dissatisfied with progress, Stalin, convinced by Vilyams, endorsed the travopolye system through the plenary meeting of the Party's Central Committee in June, 1937.[3] It was relatively inexpensive—a virtue at a time when investment was being concentrated on the development of industry at the expense of agriculture. Fields in grass could not, however, be producing the necessary grain, so that the periods in grass were reduced from those which Vilyams originally considered essential. In time, many variations became acceptable as travopolye, which made it easier for kolkhoz and sovkhoz managers to meet the official insistence on the Vilyams system, while still meeting requisitions for grain

[1] Smith (1959) Chapter 2, examines the early development of these methods.
[2] Joravsky (1967), 163 quotes authorities.
[3] Joravsky (1967), 163.

and other crops. In effect, travopolye simply came to mean a rotation including grass for at least two years. Lysenko extended the term to include one-year grass-legume mixtures.[1]

Stalin's insistence on the adoption of Vilyams' solution caused farms to turn over fields to grass in areas that were climatically and pedologically unsuitable, which offset the beneficial effects which probably accrued in northern and western districts from renewed attention to grassland farming. By 1940 the area under perennial and annual grasses had been pushed up to 16·3 million hectares, compared with 3·6 million in 1928, which represented an increase from 2·8 per cent. to 10·8 per cent. of the sown area, but even after reducing his requirements to meet the necessity of maintaining grain production, Vilyams wanted 22 per cent. of the sown area in grass. During the war years the area in grass diminished with the general reduction in cultivation which reduced the sown area in 1945 to 75 per cent. of the 1940 level.[2]

The travopolye system remained officially favoured until Stalin's death, after which the insistence on it was gradually withdrawn, though the decrees promoting it were not repealed until 1962. Khrushchev had been critical of the system and its implications before that and it probably had little influence after 1953, though the area under annual grass continued to increase. Certainly, the range of rotations in use was very wide, and the cry of the agronomists for more fertilizers and other inputs had long overshadowed the rearguard defence of an outmoded panacea.

During the war all agricultural ideals had to be subjugated to the desperate task of producing food and raw materials whenever and wherever possible, thus confirming the pre-war trend towards diversification. With the post-war reconstruction programmes, particularly when the death of Stalin made possible a fresh look at the whole situation, there was renewed attention to the possibilities of specialization. This was again associated primarily with the need to overcome the perennial problem of grain supply. With the turning towards specialization to solve this problem there was also a revival of interest in the possibilities of controlled development through the

[1] *Pravda,* 15.7.50.
[2] *Strana Sovetov za 50 let,* 128–129.

state farm, which had been out of favour since the comparative failure of the great grain-sovkhoz experiments of the 'twenties.

THE RISE OF THE STATE FARM

After the death of Stalin in March, 1953, Malenkov promised greatly increased outputs of grain, but approved only very modest shifts of investment from heavy industry to agriculture, quite inadequate to achieve the desired results. Khrushchev, though favouring greater freedom and more investment support for the collectives, was formulating his virgin lands scheme, in which specialized state farms were to play the major role of development, but for the time being was not powerful enough to overcome opposition to his ideas.

There was, however, ample support for the conversion of collectives into state farms. Malenkov, Kaganovich, Molotov, Bulganin and Shepilov were united in the belief that kolkhoz property was opposed to national property. They worked, therefore, for the concentration of productive means in the Machine-Tractor Stations, with simultaneous increase in Party control over them, reduction of commercial freedom for the collectives and increase of their exchange relationships with the state, and for an increase in state farms.[1]

Production statistics provided data to support increased reliance on state farms. By 1953 state farm averages showed distinctly better yields than collective farms for almost all major commodities. One western observer has concluded 'the stagnation of Soviet agriculture towards the end of the Stalin era was specifically due to the defects of the collective farms.'[2] It is not possible to say how much of the deficiency in the performance of the collectives was due to poorer supply of inputs but the case against them appears to be strong on the available evidence.

Malenkov lost the premiership in 1955, and Khrushchev inherited the agricultural problem which none of his predecessors had succeeded in overcoming He needed quick results and could not obtain them through the much-needed

[1] Ploss (1965), 107, quoting an item by Col. Sokolov, Radio Volga, 22.3.59, via *Ost Information*, 24.3.59, Bonn. On the start of farm conversions outside thc 'new lands' area, Ploss refers to Yu. V. Arutyunyan, *V soyuze yedinom—molot i serp* (Hammer and sickle in a unified alliance), Moscow, 1963, 26.

[2] Strauss (1969), 146.

investment in the fertilizer industry because basic chemical capacity was lacking. It was fundamentally the same situation as that which had given Vilyams his chance to convince Stalin that his system of grassland farming would enable fertility to be restored and production increased.[1] For Khrushchev, however, the solution lay in the accrued fertility of the natural grasslands of the semi-arid areas—the virgin lands. To develop this difficult environment, pioneering grain cultivation in Siberia and Kazakhstan on a huge scale, a further great upsurge in the creation of state farms was judged the best medium. Collective farms were converted into state farms in large numbers in Siberia and Kazakhstan and new sovkhozes were created. By 1958 there were 6,000 state farms, rising to 7,375 in 1960 and over 9,000 in 1963.

Between 1953 and 1964 the number of collective farm households fell by 50 per cent. in Kazakhstan, the Urals and Siberia, and there were large falls also in the European north-west. In the latter half of this period conversions significantly affected also the central black-earth region, Belorussia and the Baltic republics. At the end of the period Transcaucasia became substantially affected. Relatively small decreases occurred in the Ukraine and Kirgizia, while in other Central Asian republics and Moldavia the numbers of collectives increased.[2]

Although his virgin land and other development programmes called for increased reliance on state farms, Khrushchev was not, in principle, opposed to the collectives but rather desired to strengthen their position.

Among Khrushchev's supporters in the cause of the kolkhoz were the journalist Ivan Vinnichenko and Academician Strumilin. The latter had noted the starving of the collectives of investment funds and he supported associations to pool resources for intercollective power stations, irrigation systems and soil conservation schemes. Vinnichenko argued that an all-union association would permit the weak collectives to emulate the richer ones in instituting insurance schemes and monetary wage payments.

The liberalizing trends heralded at the Twentieth Party

[1] Discussed in Chapter 8.
[2] Strauss (1969), 181–182.

Congress in February, 1956 by Khrushchev's denunciations of Stalin and his economic mismanagement, extended to easing burdens on the members of collectives. Procurement prices were raised and criminal responsibility was removed from those who failed to work the required minimum number of days in the collective's tasks.[1] The new incentives and good climatic conditions raised agricultural production appreciably between 1956 and 1958. Khrushchev was at the height of his successes and at the Twenty-first Party Congress stressed the suitability of the kolkhoz system, while trying to strengthen it by intercollective ties and coordination.

Khrushchev's further support for the kolkhoz system at the Twenty-first Congress was followed by an intensive argument concerning the forced merger of collective and state farms. Vinnichenko produced economic evidence to show the superiority of the kolkhoz over the sovkhoz. He maintained that the peasants were disillusioned with the sovkhoz and hinted that the conversion of a kolkhoz into a sovkhoz was determined in the R.S.F.S.R. Ministry of Agriculture upon the petition of local leaders.[2] Khrushchev's powers even at this stage, however, were limited and he failed to convene the Third All-Union Congress of Collective Farmers. Meanwhile conversion of collectives into state farms continued. At the December 1959 Central Committee Plenum Khrushchev had to drop proposals for a kolkhoz centre and to agree that intercollective organizations should be restricted to the district level and only certain Repair and Technical Stations, which had succeeded the MTS in 1958, should become intercollective repair shops.

Despite these reverses, Khrushchev continued to improve the lot of the collective farmers. Agriculture was among the branches of the economy that benefited in the autumn of 1960 from increased allocation of resources permitted by the overfulfilment of industrial plans in 1959 and 1960. Terms of trade and credit for collectives were improved and taxation modified. The state sector, however, continued to expand. Union-republic ministries were created to help with the kolkhoz-sovkhoz conversions in the summer of 1960 and an

[1] Schwartz (1965), 79–80.
[2] Ploss (1965), quoting Vinnichenko, *Nash sovremennik*, 1959 (4), 174–192.

authoritarian was placed in charge of the R.S.F.S.R. Ministry of Agriculture, which handled conversions in this, the largest of the republics, but six months later the tide had turned again, he was removed from the post and the number of conversions was slowed down. It was claimed that the voting by collective farmers to petition Moscow for mergers with local state farms was entirely free but this is disputed.[1] In the event, the number of collectives decreased by 8,500 and sovkhozes increased by almost 1,000 in 1960. The share of state farms in the government's productive investment in agriculture increased from about 33 per cent. in 1957 to about 60 per cent in 1960.[2]

Khrushchev, as already noted, supported the sovkhoz as the chosen instrument of production in the new lands and around cities, where intensive dairying and market gardening were envisaged, but clearly the programme went far beyond these needs, and continued apace after his removal, though not without opposition. V. G. Venzher, who had survived attacks by Stalin.[3] continued to argue in favour of the kolkhoz, on both economic and social grounds.[4] He believed that the conversion from the form alleged to be 'lower', the kolkhoz, to the 'higher' sovkhoz was often made on doctrinaire grounds, yielding no increase in output. A sympathetic reviewer noted that collective farms were capable of developing into modern, specialized units, in effect 'collective farm factories.'[5]

The number of state farms, about 9,000 in 1963, reached nearly 12,000 in 1965, when the data for the distribution shown in Figure 1 was assembled. At that time, the state farms controlled about 560 million hectares of land as compared with the 36,900 collective farms' total of about 495 million hectares. In terms of sown land, the collectives still had the higher total—about 105 million hectares (excluding personal plots) compared with 89 million hectares in state farms. After the attainment of the position shown in the map, the rate of conversion slowed a little but at the end of 1969 the number of state farms had risen to 14,310 with a sown area

[1] Ploss (1965), 196.
[2] Walters and Judy (1967), 321.
[3] Stalin (1952), 93–104.
[4] Venzher (1966).
[5] A. Strelyany, *Radio-television*, 28.12.66, 11; reported in *C.D.S.P.*, 19 (3), 7–8.

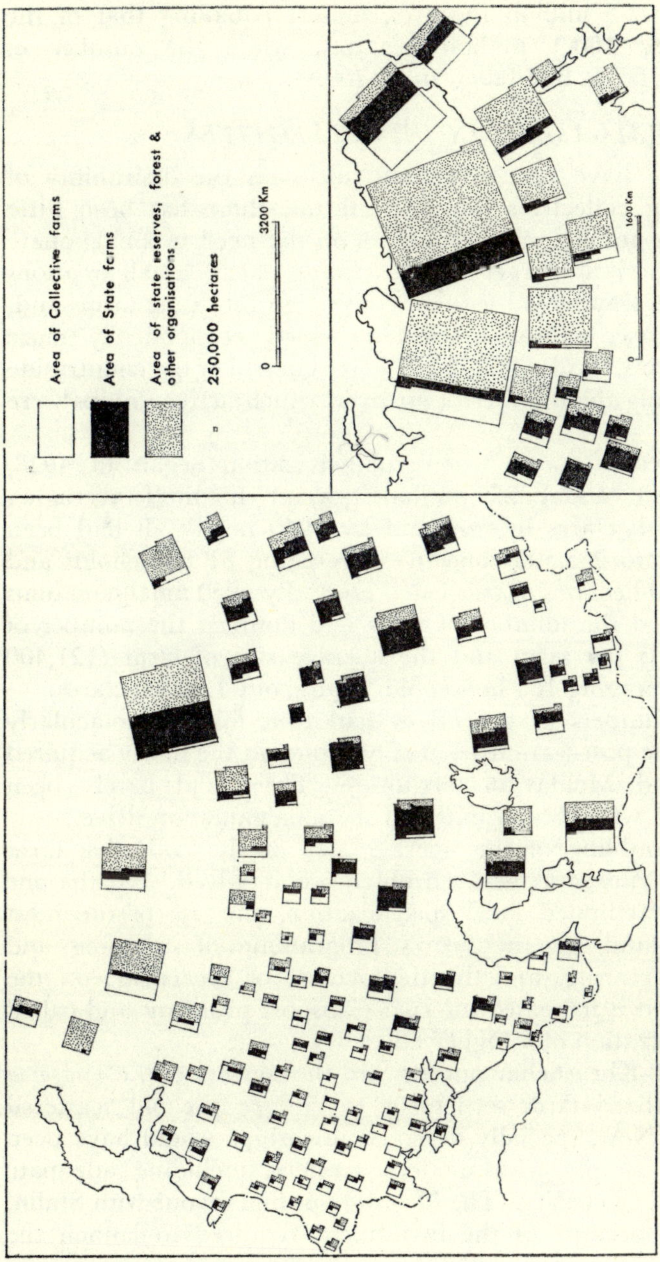

1. Regional distribution of land between collective and state farms and other users, 1965
The map shows the continued dominance of the kolkhoz in the western areas and the importance of the sovkhoz in areas of recent development in the east and also near Moscow, Leningrad and other large cities. In the north, east and in Central Asia, forest organizations and state reserves account for large areas of land
Source: *Atlas razvitiya khozyaystva i kul'tury SSSR*, 1967, 55

totalling 92·5 million hectares, almost equalling that of the collectives, 100·3 million hectares, while the number of collective farms had fallen to 34,700.[1]

THE AMALGAMATION OF COLLECTIVES

If there have been dissident voices on the desirability of converting collectives into state farms, there has been little dissension among political leaders on the need to amalgamate collectives into larger units. Some such amalgamations paved the way for subsequent conversion into state farms and, in any case, it was argued, provided economically more viable farms. It also facilitated Party control of the countryside by reducing the number of farms in which active groups were needed.

As already noted, when collectivization began in 1928, there were about 25 million peasant holdings averaging about 15 hectares in size, and by 1940 nearly all had been merged into 235,500 collectives averaging 81 households and about 500 hectares of cultivated land. By 1950 amalgamations had halved the number of units and doubled the number of households per farm and the average size of farm (121,400 farms, averaging 165 households and about 1,000 hectares).

The numbers of collectives had thus fallen spectacularly despite the post-war creation of new ones in the newly acquired Baltic and Moldavian territories. These had barely been organized when Stalin initiated the amalgamation drive.

The doubling of the average size of the collective farm probably caused as many problems as it solved. On the one hand it facilitated local specialization, the use of the most suitable land for the farm's programme of cropping and livestock rearing, and the deployment of specialists; on the other hand it increased internal transport problems and called for organization of a higher efficiency.

In 1950 Khrushchev put forward the concept of the *agrogorod* or centralized farm settlement to replace the old scattered villages. New, specially designed townships would have been easier to supply with modern services, including adequate water and electricity. The idea did not find favour with Stalin, probably because of the investment required to launch the

[1] *N.kh. SSSR 1969*, 397, 412.

scheme. Amalgamations, however, continued and by 1953 the 91,000 remaining collectives averaged over 1,400 hectares of sown land and 220 households. After Stalin's death the changes proceeded even faster and the numbers fell to 83,000 in 1956. Startling though this rate of decrease in the number of collectives was, it was left far behind between 1956 and 1962, by which year numbers were down to 39,700. The conversion and amalgamation programme has since proceeded more slowly but in 1967 there were only 36,200 collective farms, less than one-third of the 1950 figure. In 1967 there was an average of 2,814 hectares of sown land per kolkhoz, excluding personal plots. This was much less than the average sown area per sovkhoz—6,900 hectares—but compared with the average number of 617 workers on each sovkhoz, the collectives averaged 418 families each, and hence a much larger potential labour force in total numbers.[1]

The size of the collective farms resulting from amalgamations varies greatly according to region. Figure 2 shows that areas which had a high rural population in the nineteenth century—and usually also today—tend to have the largest collective farms in terms of families. Thus, the Ukraine, Moldavia, the Caucasus and central black-earth areas, Uzbekistan and Tadzhikistan all average over 500 households per collective. Most of the remaining western and southern areas have 300–500 families per farm, whereas northern and eastern areas generally have fewer families. For comparative purposes, it was necessary to use statistics which included fishing collectives for this map,[2] but these are significant only in Kamchatka, Sakhalin and other thinly-populated coastal areas.

Dissatisfaction with the working of enlarged collective farms sometimes leads to local pressure for separation by the dissatisfied brigade or brigades, but official policy opposes such tendencies. Such a movement and its rejection were reported in *Pravda* in 1966.[3] At Kabayevka in Orenburg oblast employment was provided for the workers on the collective on only 119 days in the year, and one brigade resurrected a two-year-old proposal to split off and become a separate farm. It

[1] *N.kh. SSSR 1967*, 470, 483.
[2] The sources were *N.kh. SSSR 1967*, 470 and *N.kh. RSFSR 1967*, 308–311.
[3] *Pravda*, 31.7.66, 2; *C.D.S.P.*, 18 (31) 28–29.

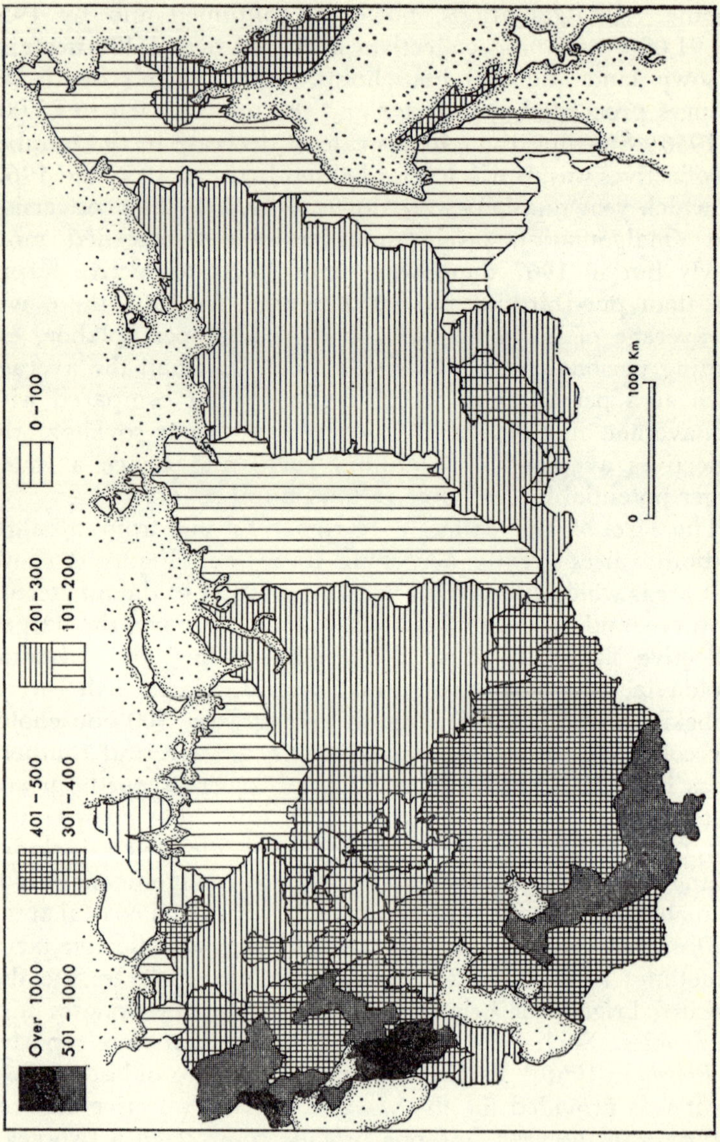

2. Average number of households per collective, 1967

worked 2,000 hectares of sown land with additional pasture and meadow land. The leaders of the movement were, however, accused by the rayon Party Committee of being saboteurs and disorganizers of production. The oblast Party

Committee criticized the local committee and obtained redress for the accused, but ruled against the division of the farm. *Pravda's* correspondent, however, was concerned at the lack of personal contact with the farmers, who still considered themselves wronged. The *Pravda* correspondent suggested that the problem could probably be solved without dividing the farm, by reorganizing the work of the brigade and introducing economic accountability.

THE ELIMINATION OF THE MTS

The conversion of collectives into state farms was only one field in which there was bitter dispute and struggle for ascendancy among groups of diverse opinions in government circles. Another concerned the abolition of the Machine-Tractor Stations, in which Khrushchev gained a notable victory.

The function of the MTS in supplying machinery for the collectives had never been discharged entirely to the satisfaction of the farm managers and committees. Complaints were common about machines not being available when required and of preferential treatment to some farms when demand for particular machinery was general. The MTS were the local bases for the Party control of the countryside[1] which did not endear them to those whose job was essentially productive and non-political. There were economic reasons, also, for dissatisfaction. The farms had to pay in kind for the work done by the MTS but there was no proper system of cost-accounting, and the payments were not related to the efficiency of the work.

Occasional suggestions before 1953 that the collectives should take over responsibility for machinery were quickly and firmly squashed by Stalin, as, for example, in answer to Sanina and Venzher who advocated sale of MTS machinery to release state finance for investment elsewhere:

> The effect would be to involve the collective farms in heavy loss and to ruin them, to undermine the mechanization

[1] See, for example, Laird (1958), 102–109. Laird concludes (p. 108) 'Quite clearly, by the beginning of 1958, the control over agricultural production had almost completely fallen into the hands of the MTS organization.'

of agriculture, and to slow-up the development of collective-farm production.[1]

'Such expenditures' argued Stalin, 'can be borne only by the state, for it, and it alone, is in the position to bear the loss involved by the scrapping of old machines and replacing them by new. His successors were, however, not as convinced as Stalin of the productive advantages of the situation. As Khrushchev gained control, he swung openly towards reducing the role of the MTS. In 1956 experiments in consolidation of MTS tractor brigades with kolkhoz field units were made. During 1957 arguments for putting machinery under the control of the collectives were developed. In May a kolkhoz in Krasnodar kray bought equipment from an MTS, and Radio Moscow reported vertical integration of managerial functions, with one man acting as both MTS director and kolkhoz chairman.[2]

Action had to wait until Khrushchev had organized sufficient support in high circles, but in November, 1957, Vinnichenko initiated public discussion by reporting the advantages discovered in the experiments in partial amalgamations already tried. In the ensuing discussion by no means all were in favour, though economic arguments for amalgamation were strong. It was claimed that one district merger cut by 30–40 per cent. the state's outlay for produce previously obtained as payment-in-kind from the collectives. Khrushchev publicly enumerated the high costs of the MTS and the advantages of selling their equipment to the collectives, and made several recommendations to speed up the transfer. *Izvestiya* reported in February 1958 the sale of an MTS and repair shop to a kolkhoz and it was claimed that farms did almost three-quarters of the 1958 spring field work with their own tractors.[3]

At the Central Committee's plenary session in December, 1958, Khrushchev reported that over 55,000 collectives (81 per cent.) had purchased machinery and 80 per cent. of the MTS had been reorganized in their attenuated form as Repair and Technical Stations (RTS), intended to handle major jobs still beyond the power of the collectives, and the

[1] Stalin (1952), 100.
[2] Summary after Ploss (1965), 105–107.
[3] Summarized from Ploss (1965).

sale of machines and fuel. The transfer, however, had still not been as extensive as he wished, and he criticized farms which had not seized their opportunities and still depended on the remaining MTS.

For the weaker collectives, the transfer involved parting with scarce capital resources, and some did not want the responsibility of owning and operating machinery, and eventually replacing it. For the more progressive, however, it narrowed the gap in resources and potential efficiency between their own operations and those of the state farms, which had always had their own machinery. Whether or not one accepts Stalin's argument that the collectives could not have afforded machinery in 1952, they were certainly generally impoverished then, even if only because they had been starved of resources under Stalin's heavy-industry programmes. Khrushchev had agreed that the collectives were not strong enough economically in earlier years to buy their own machinery and use it efficiently, and had cited the poverty of the collective farm in his native village of Kalinovka. He had said that the collective was so poor in 1945 that it had refused to accept, free of charge, horses offered by the Soviet army because it could not afford to feed and care for them.[1] By 1958, when over 2,000 million rubles was paid by the collectives for MTS machinery, there must have been a considerable improvement in their finances.

The RTS did not long survive. Even in 1958 Khrushchev was advising collectives to join together to buy or build intercollective repair shops. As experience with the RTS grew during 1959 it was argued that their handling raised prices of machinery to farms by 13–20 per cent. Before the end of 1959 all RTS units in Latvia had handed over their sales work to a Main Administration of Agricultural Trade under the Council of Ministers of the republic. RTS units in the area were renamed Machine and Soil Improvement Stations (MMS). In Tadzhikistan the RTS were abolished about the same time.[2] Within the next year or two they were generally superseded and the remaining workshops were transferred to regional departments of a new Farm Machinery

[1] Schwartz (1965), 117.
[2] Ploss (1965), 162–163.

Association (*Soyuzsel'khoztekhnika*). The farms have some choice in whether or not they take their problems to the latter. The F.M.A. has the advantages of better facilities, specialist technicians and a pool of exchange machines from which it can replace a unit brought in for repair. Reports of the efficiency of the workshops vary, some collectives reporting good and improving service, others poor service and a shortage of replacements for machines under repair. The 23rd Party Congress stated that complex repairs should always be undertaken by the F.M.A., farms dealing with minor running repairs. Between 1965 and 1967, however, over 1,700 workshops and small factories were reorganized as units of the F.M.A.[1] and the association seems to have gained increasing recognition.

Services performed by the F.M.A. are among those for which credits have been made available from the U.S.S.R. State Bank to collectives. Officers of the F.M.A. have been delegated to work at major plants manufacturing fertilizers and weed killers to improve their quality, while local F.M.A. organizations maintain railside warehouses for the storage of fertilizers, etc.[2]

The F.M.A. is equipped to apply fertilizers for farms, and lower costing than for application by farms has been reported. Thus, in Ryazan oblast, costs of applying a ton of mineral fertilizer on collective and state farms in 1965 averaged between 4·42 and 7·63 rubles, whereas the F.M.A. cost was 3·38.[3]

The Farm Machinery Association has become the object of criticism similar to that formerly levelled at the Machine-Tractor Stations. Although their control is less, they occupy a key position, controlling the supply of spare parts for machinery, electrical items, fertilizers and other necessities. Accusations have been made of spare parts being sold illicitly and of F.M.A. operators claiming for work alleged to have been undertaken for collective farms which was never done. But the collectives are in a weak position. 'Quarrelling with the Farm Machinery Association,' said one collective farm

[1] *Pravda*, 23.3.67, 2; condensed text in *C.D.S.P.* 19 (12), 26–27.
[2] *Izvestiya*, 12.3.67, 2; *C.D.S.P.* 19 (10), 38.
[3] A. Ugryumov, *Pravda*, 28.2.67, 2; *C.D.S.P.* 19, (9), 24.

chairman, 'is like spitting into the wind; they will listen to your criticism, but then choke off your supplies legally until you could scream.'[1]

The F.M.A. is charged with supplying the farm with all electrical materials, but, according to a Kurgan engineer, its offices had no electrical engineers and were sometimes incapable of merely drawing up lists of requirements, while delays in supply were inexcusable. It was said to be necessary to go searching for materials and even to engage in barter on behalf of one's farm.[2]

PERSONAL PLOT PROBLEMS

The right of the collective farmer to maintain a garden plot and keep a small number of animals of his own was written into the charter which provided the legal basis of the collective farms.[3] It has not, however, escaped attack. Personal land holdings of any kind seem to many a survival of the pre-capitalist or capitalist and bourgeois systems and an impediment to the establishment of communism. It was probably permitted at first only because the bitter resistance of the peasants to collectivization had somehow to be appeased. The personal plot represented a concession to the age-old desire of the peasant to own land and livestock of his own. Nevertheless there was, and still is, economic justification for the private plots because of the high yields extracted by the intensive labour put willingly into them. Thus, political objections are outweighed by social and economic factors and the personal plot has survived and even gained in strength over the three decades or so since it gained legal recognition.

In 1939 a decree with the purpose of establishing a minimum number of labour-days for members on the collective lands, stated bluntly some of the 'distortions' in land tenure:

> The diversification and dissipation of the collectively-held fields of collective farms for the benefit of the individual farming of the members comprehends various kinds of unlawful additions to house-and-garden plots which increase them beyond the size provided for in the Charter, either

[1] G. Radov, *Literaturnaya gazeta*, 18.10.66, 2–3; *C.D.S.P.* 18 (47), 15.
[2] S. Podkorytov, *Pravda*, 8.12.66, 2; *C.D.S.P.* 18 (49), 28–29.
[3] See footnote 2, page 7.

> on the pretext of feigned separations of families, where a household fraudulently obtains an additional house-and-garden plot for members of the family who pretend to be separated, or by direct allocation of house-and-garden plots to the collective farmers at the expense of the collectively-held fields of the collective farms. . . .
>
> In a number of collective farms, the practice is in reality to transform the house-and-garden plot into the private property of the household, so that not the collective farm but the individual member of the collective farm disposes of it at his own discretion, i.e. rents it or retains the plot for his own use, although he himself does not work in the collective farm. . . . (Such actions) tend to deprive farming on house-and-garden plots of the character of an auxiliary source of income, and such farming is sometimes transformed into the main source of income of a collective farmer.[1]

New rules made it more difficult for the peasants to keep private livestock and their numbers fell sharply as many had to be handed over to the collectives.[2]

During the war years the kolkhoz system was inevitably dislocated, with varying effects in time and place according to whether or not the area was occupied by the Germans and the degree of control that could be exerted by the Soviet authorities. Many peasants undoubtedly hoped that after the war they would have greater freedom for private farming, but the new legislation of 1946 ordered restoration to the collectives of property and livestock illegally seized by peasants.[3]

Nikita Khrushchev, then in a commanding position in the Ukraine, however, believed in maximizing incentives and reducing central control in the agricultural industry. Khrushchev strongly favoured the retention of the personal plot, and influenced others. A. A. Andreyev, who was put in charge of the ill-fated Council on Kolkhoz Affairs,[4] persuaded the C.P.S.U. Central Committee Plenum in February, 1947, that each kolkhoz household should be allowed its own cow for

[1] Decree of the Central Committee of the Communist Party and the Council of People's Commissars, 27.5.39, trans. Gsovski (1948–49), vol. 2, 475–476.

[2] Jasny (1949), 341, 356.

[3] Decree of the Central Committee of the Communist Party and the Council of Ministers, 19.9.46, trans. Gsovski (1948–49), vol. 2, 487–497, see also Meisel and Kozera (1953), 388–394.

[4] Formed in October 1946, the Council was abolished by Stalin who disapproved of Andreyev's policies.

private use.[1] Similar empirical measures were urged by N. A. Voznesenskiy, who, as an economist, believed in monetary incentives and cost-accounting. Of these three, however, only Khrushchev held on to office through Stalin's purges and for a time only subordinate to a new Party chief in the Ukraine. Khrushchev had had many brushes with Stalin, often arising from representations the Ukraine leader made on behalf of the farmers. After complaining about the high tax on private apple orchards, Khrushchev later recalled, Stalin 'replied that I was a *narodnik* (peasant socialist), that mine was the *narodnik* approach, and that I had lost the proletarian class instinct.'[2]

Khrushchev was also frequently opposed to Malenkov, but Malenkov, when he became premier, accepted the necessity for encouraging consumer goods industries and agriculture. In August 1953 he announced, along with incentives for increased production, that state hostility to personal plots would end and quotas for compulsory deliveries of garden produce be cut substantially.[3] A month later Khrushchev stressed again the desirability for workers to own livestock as private property.[4] This policy was partly reversed in March 1956 by an order which required collectives to control the size of plots according to the household's contribution to the collective.[5] Those that made no contribution to the collective work were supposed to be deprived of their plots, and there was a reduction in personal crops and livestock. Restrictions on the personal plot were temporarily hardened and it was made increasingly difficult for kolkhoz workers to keep their own cows. Then, however, followed another period of easing of restrictions, and compulsory deliveries from private plots were abolished from 1st January, 1958. When Brezhnev and Kosygin succeeded Khrushchev they further lifted restrictions. It was expressly stated that plots and livestock could be held not only by members of collective farms, but by single, aged and disabled members of collectives

[1] *Pravda*, 28.2.47 and 7.3.47.
[2] *Kommunist*, 1957 (12), 12–13, quoted by Ploss (1965), 34.
[3] *Pravda*, 9.8.53. For the extent of the reforms in taxation of personal plots see Wronski (1957), 201ff.
[4] *Pravda*, 15.9.53.
[5] *Pravda*, 10.3.56.

reorganized into state farms, and by industrial workers, and owners were to be helped to acquire livestock and to store fodder. Taxes and restriction on the size of plot were eased. Allocation of land by the collectives for additional pastures for personal livestock was also encouraged.[1]

Relaxation of pressure against the personal plots has undoubtedly been associated with the retarded production experienced by both collective and state farms. It has been said that improving conditions for collective and state farm workers would eventually make the personal plot 'unwanted', but clearly this is far from realization yet. Intensive production of potatoes, vegetables and fruit for home consumption or sale, and grain and roots for his own cattle, pigs and poultry remain vitally important to the worker and to the national economy. In 1969, more than one-third of the total production of meat, milk and vegetables, 56 per cent. of the eggs and 67 per cent. of the potatoes produced in the U.S.S.R. came from these plots, which amounted to under 3 per cent. of the land under cultivation.[2] Previously the proportion of these commodities so produced had been higher but the trend has been for these proportions to diminish with improved output from the collective and state farms. In absolute quantities, however, the yield from the personal plots remains remarkably steady, partly because roughly three-quarters of it is consumed on the holding and effort can be directed into providing subsistence or market crops according to the needs of the family and the public-sector supply situation.[3]

In state purchasing, the personal plots matter much less than in total production, but in 1969 they provided 21 per cent. of the potatoes and 12 per cent. of the eggs purchased, and 14 per cent. of the wool, besides smaller proportions of vegetables, meat and poultry.[4]

These figures make clear the continuing importance of the personal plot and livestock to both individual and the state. In comparing the high yield of these small areas with the disappointing yields obtained from agriculture in the Soviet Union in general, some redressing of the balance is necessary.

[1] *Pravda*, 6.11.64 and 7.11.64.
[2] *N.kh. SSSR 1969*, 295.
[3] Lovell (1969).
[4] *N.kh. SSSR 1969*, 300.

Small plots, worked intensively, normally yield more highly than the broad acres under mechanized cultivation, particularly where large quantities of fertilizers have not been generally available. For the kitchen garden, refuse rich in nutrients and manure from the domestic animals supply the fertilizer. Work by the able-bodied men in off-duty hours can be supported liberally by the elderly and the young. There is little potential in the garden plots and their single cows, however, for the technological improvements that are slowly beginning to benefit the collectives.

Inevitably, private plots compete with the collectives, not only for labour, but also for other resources. In the summer of 1967, *Pravda* carried a report that state farm workers were complaining that they were unable to obtain enough fodder for their personal livestock because of high demands from the government purchaser.[1]

Furthermore, higher incomes from the collective operations somewhat reduce the appeal of the personal plot and encourage better work on the collective lands. A *Pravda* correspondent was shown a Tambov farm: ' "Look, on one side of the road is the communal collective farm land, on the other the personal plots. Guess which is which." On both sides the soil was beautifully tilled, levelled and clean, and I could not guess, in the context of old conceptions.'[2]

The present writer can better this by reporting that on a farm in Stavropol kray in 1968, all the collective fields and orchards were immaculately cultivated while many of the private plots and fruit trees looked unkempt and some owners had given up keeping their individual cows. Presumably, on this farm at least, incomes had reached the point where some peasants preferred increased leisure to raising more personal produce. This farm, however, may have been atypical in being particularly well managed and at the same time somewhat distant from kolkhoz markets.

THE PRICING OF AGRICULTURAL PRODUCTS

A major overhaul of the prices paid to collective farms followed the abolition of the MTS. Until then several different price schedules were in operation. Prices differed for the same

[1] *Pravda*, 15.7.67, 2.

[2] A. Volkov, *Pravda*, 27.6.67, 2; *C.D.S.P.* 19 (26), 37.

commodity sold to the state as part of the compulsory delivery quota, as a sale above the quota, or to the MTS as payment in kind. Bonuses could be earned on overfulfilled plans, which placed a premium on obtaining a low plan and also on the better-endowed farms. Khrushchev cited collectives in areas producing cotton, flax, hemp, citrus fruits and dairy products as particularly able to earn large incomes when they were able to qualify for bonuses. Average and weak farms had to deliver all or nearly all their output at the low compulsory delivery prices, which did not always cover production costs.[1]

Procurement prices had been increased from time to time, of course, and differential movements made to help backward sectors. In 1953–54, for example, adjustments of tax and procurement prices were designed to benefit producers of animal products, vegetables and grain.[2] By 1959 procurement prices had tripled since 1952[3] and total peasant income in cash and kind probably doubled in step with the farm incomes.[4] Prices, however, were still kept artificially low to keep down costs of food in the cities, the multiple-price system continued and much kolkhoz output had to be sold at below the cost of production.

The 1958 price reform aimed at redistributing expenditure in procurement, without increasing the total previously paid to farms and MTS stations for produce purchased from them. One price only was fixed for each crop, varying according to area, this price being above that previously paid for compulsory deliveries but below that paid for bonus quantities. Compulsory deliveries remained but the higher prices paid for these quantities ensured better incomes for the poorer farms, while moderating the profits of the more fortunate. There was also considerable adjustment in the relative prices for different products, those for all major livestock products being substantially raised, those for some arable crops slightly reduced.[5]

The terms of trade for collectives were improved again in 1961 by reductions of prices for tractors, vehicles, machinery

[1] *Pravda*, 21.6.58.
[2] Davies (1955), 61.
[3] Volin (1967), 7.
[4] V. Khlebnikov, *Voprosy ekonomiki* (7), 1962, 49, estimated that total peasant income doubled between 1953 and 1958, quoted by Ploss (1965), 147.
[5] Schwartz (1965), Table 14.

and fuel, income tax concessions and improved credit facilities.[1] Many farms were enabled to distribute payments entirely in cash, though an editorial article in *Sovetskaya Rossiya* on January 25, 1962, noted that only one-quarter of the collectives in the R.S.F.S.R. did so, and advocated the extension of the practice, which was also urged by other authorities.[2] In 1962 prices for livestock products were again raised, by an average of 35 per cent., involving higher prices to consumers, because production had to be stimulated and many farms were still losing money on their produce.[3]

The areal variation in prices henceforth paid for any crop introduced a new dimension into pricing practice for agricultural products in the U.S.S.R.—the recognition that in the absence of land rents some method was needed to increase rewards to farms on inferior land and to reduce the excess profits accruing to those with exceptionally fertile soils and favourable climates. Although there had long been academic concern with problems of agricultural regionalization, dating back to the eighteenth century,[4] hitherto the well-known variations in regional potential had not been applied to pricing systems. The collectivization movement was accompanied by efforts to reconstruct the geographical distribution of agricultural production to bring it more into line with the potential inherent in the physical characteristics of different areas, but maps of agricultural regions intended to promote this movement did not gain official favour.[5] Stalin then announced a policy of agricultural diversification to give every industrial region its local source of foodstuffs and reduce the burden on transport. Before the Second World War steps had been taken to minimize the transport of potatoes and to reserve areas around cities for vegetable and milk production. Specialization around cities remains standard policy, but the policy to develop products regionally favoured by climatic and soil conditions was announced at the Twentieth Party Congress in 1956.[6]

[1] A. Pelshe, *Pravda*, 14.1.61; *C.D.S.P.* 13 (5), 8–10.
[2] Ploss (1965), 243–244.
[3] *Pravda*, 1.6.62; *C.D.S.P.* 14, (20), 3–5.
[4] See Chapter 10.
[5] Jackson (1961).
[6] Zaltsman (1956), 1960.

The zones delimited for pricing purposes did not accord in detail with natural conditions but grouped oblasts and other administrative areas with broadly similar conditions for given products. Particularly favoured zones were allocated low prices, higher prices for the commodity concerned being fixed for less favoured areas. For areas with a very low suitability for the particular product no price was set, implying a directive that such an area should concentrate on products for which it was more suited. This price variation did not fully offset the variations in production cost, so that areas particularly suited to wheat, for example, continued to derive higher profits from the crop than areas less suited. There was, therefore, some subsidizing of the poorer areas but not enough to eliminate natural advantages between zones. Within zones, variations in climate and soils also produced considerable local differences in profitability. The failure of the zones to accord with natural conditions and their large size have led to criticism,[1] but they have given useful assistance to farms in regions not especially well endowed climatically to maintain a range of products, while not discouraging closer adherence to production patterns appropriate to the natural conditions. To define zones in great detail would be extremely difficult. It would require a fully detailed land inventory for every crop, and to compensate fully all handicaps to production would discourage prudent specialization in a way that no other country has done with either subsidy or support schemes. The fault of the Soviet leadership lies in this case mainly in claiming for the scheme a greater equalization of farm incomes than could ever have been intended.

In 1965, Brezhnev, at the Plenum of the Central Committee, proposed some amendments to Khrushchev's zonal pricing scheme, including a finer network of regions, approximating somewhat more closely to production conditions. Thus, in the R.S.F.S.R., the number of price zones for wheat was increased from eight to twelve. This increase did not eliminate sharp changes in prices along administrative boundaries where there appear to be no changes in production costs.[2] Substantial price increases were authorized, particularly benefiting some

[1] See, for example, Jensen (1969), esp. 331–335, and Bornstein (1969).
[2] Jensen (1969), 342.

high-cost areas. In the Polesye, northern Russia and the Baltic republics, prices for wheat and rye were increased by 45 rubles per ton, in low-cost areas by only five rubles per ton or less.[1] Thus, the production advantages of the more fortunate farms were further reduced, but still not sufficiently to eliminate completely regional advantages of climate, soils and accessibility.

In addition to the regional price variation, there are also variations for quality, while deliveries above the procurement plan earn substantial premiums, amounting to 50 per cent. of the zonal price for bread grains and from 10 to 100 per cent. of the zonal prices for livestock products. Farms accordingly benefit doubly in good years but suffer sharp contractions of income in bad years. The production of many commodities has undoubtedly benefited from these price incentives. The considerable scope for improvement was shown by the report that in 1965 over 25 per cent. of the collective farms and 50 per cent. of the state farms (which were paid at lower rates than collectives for their deliveries because they were state-financed) failed to cover their costs.[2] 1965, however, was a poor year for production, especially of grain, and 1966 and 1967 showed crop production indices some 12 per cent. higher,[3] which should have benefited income. The prices set for the period 1965–70 seemed likely to cover costs of production for the average farm, even in some high-cost areas.[4]

MARKETING

The compulsory deliveries to the state procurement system having been met, farms have some degree of choice in the disposal of other saleable produce. The freedom to trade in the kolkhoz markets has been greatly valued by the farmers and consumers alike. Legally constituted in 1932, these markets are the descendants of the fairs of former times and exist in most towns. In the post-war period their turnover amounted to about 15 per cent. of total retail trade.[5] Stalin

[1] Jensen (1969), 340.
[2] Khlebnikov (1967), 10–11, quoted by Jensen (1969), 340.
[3] *N.kh. SSSR 1967*, 328.
[4] Walters and Judy (1967). For improvement in the previously very low returns to dairying, see Strauss (1970), 279.
[5] Whitman (1956), 391–399.

disliked the kolkhoz market places and envisaged their disappearance as kolkhoz property was converted into public property, with products-exchange replacing commodity circulation based on cash transactions.[1] After Stalin's death, policy turned to state competition with the kolkhoz markets through higher procurement prices and development of co-operative buying on commision.[2]

The degree of independence a collective should enjoy in the disposal of its produce and its organization are subjects of considerable dispute. The liberal view is expressed in Venzher's definition of a kolkhoz:

> The collective farm is a socialist agricultural enterprise based on social ownership of the means of production and the collective labour of its members, *independently planning* the development of their communal economy, bearing full material responsibility for the results of their economic activity, *freely disposing* of the products created by the labour of the members, and entering into economic ties with the state production sector of the national economy and with cooperatives *exclusively* on the basis of commodity-monetary relations with obligatory, absolute observance of the principle of equivalence.[3] (my emphasis).

Venzher further claimed for kolkhozes that 'a system of free sale of the output best suits the nature of collective farm production.' Such views were strongly attacked by senior members of the Lenin All-Union Academy of Agricultural Sciences, the Agricultural Economics Department of Moscow University and others.[4] They asserted that, contrary to Venzher's view, mass marketing to a wholesale purchaser best fits the large-scale nature of collective farm production. Therefore, they argue, the task is to work for perfecting the system of planning the procurements of agricultural produce. The system, they admit, must more fully take into account the specialization of farms and determine the products to be sold accordingly, making use of the price mechanism, but apparently subordinating it to administrative decision.

[1] Stalin (1952), 76.
[2] Whitman (1956), 401–404.
[3] Venzher (1966), 141–142.
[4] S. Kolesnev, M. Sokolov and I. Suslov, *Sel'skaya zhizn'*, 22.9.66, 2–4; *C.D.S.P.* 18, (37), 7–10. Stalin's attacks on Venzher were noted earlier.

Following complaints in 1966 about collective farm markets, including lack of use at Krasnodar and high prices at Novosibirsk and Krasnoyarsk, the U.S.S.R. Ministry of Trade commented on the facilities at these markets and the trading practices they should observe.[1] It was admitted that in some markets, where lists of maximum sales prices were posted, pressure was being exerted on traders to adhere to these prices, which led to trade outside the market. On the other hand, at Tambov, a trade service bureau had been established to receive produce from the collective farmers and to remunerate them according to sales, resulting in an increase in the market's trade and falls in meat prices.

It was emphasized that the U.S.S.R. Council of Ministers had directed that the entire income of kolkhoz markets was to be used for the improvement of facilities at the markets. Such facilities included equipment, special clothing, storage rooms and refrigerators, repair and catering enterprises and the like. Hostels to accommodate visiting farmers had been built for many of these markets. Unfortunately, however, the supply of construction materials for these projects was given low priority. In the Ukraine, independent inspectors had been appointed, collective farms were advised of market prices and a competition for the best market had been launched.

The amount of trade handled by the kolkhoz markets seems to have declined greatly with the improved prices offered by the state. It appears now to be only about 2–3 per cent. of the total, though varying from area to area. The inability to purchase large quantities because of lack of storage and refrigeration equipment hinders the markets from handling the business of the big collective farms. Many enterprising farms seek direct sales to plants such as meat-packing combines and flour mills. Some deal with distant trade agents, disposing of their surpluses at legal but profitable prices in return for guaranteeing the quality of their goods. Crimean apples and grapes are sold direct to Bratsk, Kamchatka and the Baltic republics. 'They are serious customers,' said a collective farm chairman.[2]

Not all buyers are held so highly. Some buy produce as

[1] N. Kuzmenkov, *Pravda*, 5.10.66, 3; *C.D.S.P.* 18 (40), 33–34.
[2] G. Radov, *Literaturnaya gazeta*, 1.11.66, 1–2; *C.D.S.P.* 18 (47), 18–19.

sub-standard—making this grading of a percentage of the goods a condition for handling them—and then selling them at first grade prices. Frequently, a commodity deteriorates because the handling agency will not or cannot collect it from the farm quickly enough. At best, this results in fatstock passing their prime before being butchered, but often the results are more serious, as with fruit and vegetables not moved to market quickly enough. Direct trading is one alternative for the collective, processing on the farm is sometimes another.

As production increases, cries for more processing plants and better utilization of existing processing works become more common. In the discussion in the Supreme Soviet on the plan and budget for 1967, a deputy from Belorussia stated that the processing capacity in the dairy, meat and flax industries was lagging behind production,[1] from Vitebsk oblast came a complaint of the lag in the growth of capacities for processing sausage, cheese and whole-milk products,[2] while the Moldavian representative said that in his republic the shortage of canning and other processing plants was acute, and another sugar refinery (to make ten in all) was needed. For fruit, processing facilities in this area were said to be virtually non-existent and produce had had to be sent to plants over 200 kms. away, even though the latter had no surplus capacity. The U.S.S.R. Ministry of the Food Industry was accused of dropping a plan for a new cannery without the republic's agreement or knowledge.[3]

In concluding remarks on the debate on the plan, however, it was stated that the necessary production base for canneries had not been developed in many areas in the Russian republic, the Ukraine, Georgia and Azerbaydzhan and only two-thirds of the capacities of the canneries were utilized. It was recognized, however, that processing capacities must be increased and plants on collective and state farms to process their own products had a large role to play.[4]

Changing patterns of demand as well as increasing supplies have been reasons for requiring growth in processing capacity.

[1] *Izvestiya,* 18.12.66, 5; *C.D.S.P.* 19 (4), 17.
[2] *Izvestiya,* 17.12.66, 3; *C.D.S.P.* 19 (2), 21.
[3] *Izvestiya,* 18.12.66, 2; *C.D.S.P.* 19 (3), 13.
[4] *Izvestiya,* 20.12.66, 3; *C.D.S.P.* 19 (6), 18.

Kaluga oblast increased its production of milk by nearly 50 per cent. in only two years but found itself in difficulties because of a fall in demand for milk from Moscow, which was drawing supplies increasingly from Moscow oblast. As further increases in milk production were planned, Kaluga needed at least ten new dairy plants, capable of processing 50–100 tons of milk per shift.[1]

TRANSPORT

The marketing of agricultural produce in the Soviet Union depends on the ability of the transport network to carry produce over much greater distances than are usual in most other countries. Comparable and greater distances are spanned by the trade links between Great Britain and her overseas suppliers but these links are almost entirely by sea, whereas the Soviet long-distance freight links are mainly by railway, with limited quantities moving by inland waterways, sea and air. Road transport is used mainly for short hauls but is of great importance to farms because by it they can convey produce to local markets by their own vehicles at minimum cost.

The greatest problem in food commodity transport facing the Soviet government has always been that of grain movement. Before the revolution the export of wheat and other grains stimulated construction of lines from the grain-growing areas in European Russia and the Ukraine to ports on the Baltic Sea and, later, the Black Sea. In 1875, grain accounted for 42 per cent. of rail freight traffic.[2] With the expansion of industrial traffic the contribution of grain and flour fell to 13 per cent. by 1913 but 20 million tons were carried in that year.[3] After the war and the ensuing civil war the transport system was in a deplorable state and reconstruction had to be undertaken, but, with first priority being given to basic industry, transport did not receive the amount of investment needed though some new lines were constructed. Soon, however, it became clear that the railways could not cope with the traffic in grain and other basic commodities, and the problems of food supply associated with the introduction of

[1] *Izvestiya*, 19.12.66, 4; *C.D.S.P.* 19 (5), 16.
[2] Westwood (1964), 78.
[3] Westwood (1964), 138.

collectivization were intensified. By 1934 Stalin, aware of the seriousness of the problem, introduced measures to put the railways into better order. Production of new locomotives and rolling stock was increased and efficiency of working was greatly improved, notably in turnround of wagons and consequently in their utilization.[1] Average turnround time was reported to have fallen from 10·6 days in 1928 to 7 days in 1937. There has since been less scope for increasing efficiency in this direction, but an average of under 5½ days was reported for each year from 1965 to 1967.[2]

The continuing increase in the burden placed on the railways is shown by the amount of grain movement by rail in recent years, amounting to 80–90 milliard ton–kms. per annum during the 'sixties until 1968 when it surpassed 106 milliard compared with 30·9 milliard in 1950[3]. Both factors here—length of haul and weight of grain moved—have shown a tendency to increase. The increase in length of haul has been associated with regional specialization and development of new lands, especially in the 'fifties. The average haul per ton of grain rose from 795 km. in 1950 to 1,152 in 1960, (almost exactly repeating the wartime rise to 1,150 km. in 1945). It fell to 922 by 1966, but rose again to 1,075 km. in 1968.[4] Some of the variation is, of course, connected with the different experiences with harvests in the various regions.

Statistics showing tonnage of grain dispatched reveal a correlation between the poor harvests in Kazakhstan and much reduced shipments in 1965. In the same year the outward movement of grain from the Ukraine was high following good harvests in both 1964 and 1965.[5] However, the biggest weight of traffic is to and from stations in the R.S.F.S.R. On balance, the R.S.F.S.R. has usually been a net importer of grain but it was a net exporter to other parts of the Union in 1965 and in 1967. Record levels of grain harvests in the R.S.F.S.R. in 1964 and 1966 (respectively 87 and 100 million tons) presumably explained the heavy movements of 1965 and 1967.[6]

[1] Hunter (1957), Williams (1962).
[2] *N.kh. SSSR 1968*, 467.
[3] *N.kh. SSSR 1965*, 461; *1968*, 463.
[4] *N.kh. SSSR 1965*, 462; *1968*, 464.
[5] *N.kh. SSSR 1965*, 312, 313, 472.
[6] *N.kh. SSSR 1967*, 529.

Important in the R.S.F.S.R. grain movements are those from the north Caucasus plains to the consuming areas, and from western and central Siberia to the Far East, opposite to the main flow westward. The Baltic republics, Belorussia and the Central Asian republics are net importers of grain, so adding to the burden of the railways of the Russian republic as well as of the peripheral republics.

Sea transport now conveys more of the grain than formerly, the tonnage moved rising from 1·2 million in 1950 to 6·7 million tons in 1965 and 6·2 million tons in 1966 and 1967,[1] but the relief to the railways is only marginal. The same may be said of grain transport by inland waterways which was of similar volume, averaging 6 million tons in 1965–67—relatively little changed from the 1940 figure of 5·2 million tons.[2]

In general, rising production of all agricultural produce and the much greater needs of the enlarged cities has meant increased demand for transport. Specialized vehicles have been in increasing demand for meat and dairy produce, and refrigerated vans are now used in large numbers on Soviet railways. Some perishable fruit is transported by aircraft, not only under official arrangements but by enterprising collective farmers who find it worth the cost of air travel (which is low by European standards) to take the more valuable produce to city markets.

The large number of points from which traffic originates and the seasonal flow of much agricultural produce also place heavy burdens on the carriers. The build-up of stocks of other commodities has to be begun early in the year owing to the demands of agriculture in the harvest months and, preceding the harvest, the granaries have to be cleared by July or mid-August, according to region.[3] The farms prefer to move the grain as early after harvesting as possible and clamour for wagons but few grain-producing farms would have other seasonal commodities to help smooth the demand for wagons. Many grain farms also keep commercial livestock herds but this is no help to the railways. Milk requires special tank vehicles and is therefore an entirely separate problem to the

[1] *N.kh. SSSR 1967*, 533.
[2] *N.kh. SSSR 1967*, 536.
[3] Hutchings (1971), 141–142.

railways, and livestock also require specialized vehicles. Complaints from the farms of unsatisfactory transport services are common. The farm loses if produce deteriorates while awaiting movement or during unnecessarily prolonged transport. In some areas railway lines are great distances apart so that farms have a long haul to the railway, often on poor, unsealed roads.

Road transport would often be much preferred by farms, availability and costs permitting, to rail movement because of its flexibility. In fact, however, the farms have little choice in the matter. Roads in the U.S.S.R. are notoriously poor and inadequate for the tasks they have to perform and traffic is accordingly forced on to the railways wherever possible. One resulting merit of this policy is that railway rates are low, but an undesirable lack of flexibility of transport also results, and the consignor has little choice of medium of transport. The mileage of hard-surfaced roads in the U.S.S.R. as a whole has been nearly trebled since the Second World War, but even the best roads, except in and around the major cities, fall far short of western European standards. Surfaced carriageways are barely adequate for two streams of traffic, though usually unsurfaced strips are available for emergency use on each side of the highway. In most areas producing grain and sugar beet, sealed roads form a small proportion of the total, for example 4·2 per cent. in Volgograd oblast, 6·5 per cent in Saratov oblast, 8·4 per cent. in Belgorod oblast. In the Russian republic as a whole 78 per cent. of the roads are gravel-surfaced.[1]

Road traffic is greatly hindered by the long winter, with snow during four to eight months of the year, and floods and boggy conditions during the thaw. In winter, heavy vehicles can travel over frozen ground and on ice, but delays and maintenance requirements result in high costs. Some areas are rendered accessible by 'winter roads', which, in summer, are impassable because of boggy conditions. The average road haul (all types of freight) remains low, only 14·6 km. in 1968, confirming the short-distance use of the road vehicles, but even this was one-quarter more than in 1950.[2]

[1] A. Nikolayev, *Izvestiya*, 26.10.66, 3; *C.D.S.P.* 18 (43), 20.
[2] *N.kh. SSSR 1968*, 487.

Although the average length of haul has risen only modestly, the increased number of vehicles has resulted in an increase of goods transported by road from 20·1 milliard ton-kilometres in 1950 to 187·1 milliard in 1968.[1] In 1950, road transport accounted for about 3 per cent. of ton-kilometres of freight moved, but in 1967 its share of nearly five times as much total freight movement had increased to 5·5 per cent.[2]

These figures apply to total road haulage. Separate figures are not available for agriculture, but the number of vehicles in use in agriculture rose from 283,000 in 1950 to 982,000 in 1965 and 1,097,000 in 1968.[3]

The cost of road transport remains a barrier to its utilization even where adequate vehicles are obtainable. Railway rates are kept low both because of the heavy usage and because of government policy. Hence, costs of road transport per ton-kilometre are reported as nearly twenty-five times higher than rail.[4] This may, however, be exaggerated because of the different types of load constituting most of the movement on the railways—coal, iron ore, petroleum, grain, for instance, all cheap to move in bulk.

TABLE 3

COMPARATIVE COSTS OF VARIOUS MEANS OF TRANSPORT FOR AGRICULTURAL PRODUCE IN NOVOSIBIRSK OBLAST, 1956–57

Means of transport	*Route*	*Kms.*	*Kopeks per ton–km.* *Potatoes*	*Vegetables*	*Milk*
Rail	Bolotnoye—Novosibirsk	125	14·8	20·5	34·8
Rail	Chulym—Novosibirsk	122	16·8	23·1	37·7
Rail	Cherepanovo—Novosibirsk	109	22·1	30·4	51·0
Road	(gravel-surfaced)	57	57·1	71·4	71·0
Road	(tar-sealed)	57	27·8	32·5	34·9

Source: Sidorov (1958), tables 1 and 3.

Transport costs reported in the regions by local administrations and individual farms confirm the high cost of road transport. An example from Novosibirsk oblast compared the cost of transport on selected routes by rail and road for

[1] *N.kh. SSSR 1968,* 488.
[2] *N.kh. SSSR 1968,* 459.
[3] *N.kh. SSSR 1965,* 395; *1968,* 415.
[4] *N.kh. SSSR 1968,* 461.

potatoes, vegetables and milk, each a product with a large market in Novosibirsk.

From the above table the following conclusions may be drawn:

(1) With the costing used, allowing fully for time of labour and vehicles and with rail charges presumably as fixed by the government, rail transport was cheaper for all commodities.

(2) Rail rates varied according to route, the haul from Cherepanovo on the Barnaul line being more expensive, especially for milk transport, than the main Trans-Siberian line, on which Bolotnoye and Chulym lie respectively east and west.

(3) Costs of transport on sealed roads are less than half costs on unsealed roads.

(4) Even on sealed roads, road transport costs more than rail transport, except for milk, for which quoted rates refer to special vehicles. The road costs are, however, quoted for shorter distances than the rail costs. Further details show road transport cheaper over some short routes (e.g. Iskitim, at 56 kms. on the Barnaul line) but remaining dearer over others, e.g., Kochenevo-Novosibirsk, 50 kms.

At the time of this enquiry the agriculturalists considered costs of transport to Novosibirsk to be excessively high in relation to production costs. One sovkhoz near the city recorded the following relationships:

TABLE 4

COSTS OF PRODUCTION AND TRANSPORT OF PRODUCE TO NOVOSIBIRSK, OB SOVKHOZ NO. 1, 1957

Rubles per centner (100 *kg.*)

	Grain	*Potatoes*	*Vegetables*	*Milk*
Cost at sovkhoz	37·3	19·5	27·7	123·7
Transport by road	1·5	2·0	2·3	2·9
Transport cost as percentage of cost at sovkhoz	4·0	10·0	8·5	2·5

Source: Sidorov (1958), 415.

While for farms near the city costs of transport of produce are from 2 to 10 per cent. of production costs, for farms further from the market transport costs are as much as four times these percentages, or even higher. Thus, the delivered cost of potatoes and vegetables, transported 50–70 kms. over unsealed roads, is one-third to one-half above the farm cost.[1]

ECONOMIC ACCOUNTABILITY

In March, 1965, the decision to transfer state farms to full economic accountability was announced at the Plenary Session of the C.P.S.U. Central Committee. This was, however, difficult to achieve and eighteen months later there was still confusion and discussion about what was implied by 'economic accountability'.

Successful farms favoured full accountability and responsibility for their own funds. One farm, located in the dry Kalmyk steppes, forecast a profit of about 800,000 rubles after the good harvest of 1966, compared with a planned figure of 384,000. Most of the planned profit had to be passed on to the state, the farm getting at best 12 per cent. of it, probably only 4 per cent. but the farm expected to keep half of the above-plan profit. 200,000 rubles had been advanced from the state budget, so that had the farm borrowed that amount from a bank it could easily have repaid it, and had it then been allowed to retain most of its profits it would have been very much better off. This was in a good year, however, and in 1965 the profit of the same farm had been only 18,000 rubles.

Asked what indices should be presented to the farm in the plan for it, the director of this farm answered, 'The wage fund and the sales volume in rubles.' 'But what if only one index confronted you?' he was then asked, 'Let us say, you have to deduct a definite percentage of your profit for the state budget, and that's all?' Then, the director said, they would have to review the whole structure of the farm. They were required by the plan to produce eggs, wool, grain, meat and some milk. With complete accountability, they would stop producing milk, which cost them 40 rubles a ton. Milk yields were only 1,000 litres per cow, the cows being bred for beef. Similarly with eggs, which, like milk, could be more economically produced on specialized farms. The farm in question

[1] Sidorov (1958), 415.

would be left with wool as its basic product, meat and grain as subsidiaries.[1]

At this time it was pointed out by G. Kutuzov, an official of the U.S.S.R. Ministry of Finance, that state farms made a profit of about 1,500 million rubles a year, but fell about 3,500 million rubles a year short of covering all their capital investments. This deficit was allocated from the budget. There was doubt about what would happen under full economic accountability, so it seemed best to transfer first to the new system the farms that appeared to be self-supporting, and investigations along these lines were planned in Moldavia, Estonia and Voronezh oblast.[2]

Kutuzov's statement was soon taken up by *Izvestiya*. F. Tabeyev, First Secretary of the Tatar A.S.S.R. Party Committee, reported on the transition of the Northern State Farm, created on 'some of the republic's worst land', in 1957 from several of the weakest artels. It operated at a huge deficit until transferred to partial economic accountability in 1962. Budget financing continued but profitability became the aim. After numerous reforms, including the wages and incentives systems, profits were achieved and mounted in successive years. After deducting budget allocations, the farm operated at a net profit over the five years 1962–66. Given the additional freedom of selling their produce at the same prices as apply to collective farms and retaining most of their profits for reinvestment, this state farm would now appear economically viable although weak at the time of transition.[3]

Another state farm had good reason for dissatisfaction with the results of its transforming losses into profitability. After being tentatively set at 96,000 rubles, the planned profit was revised four times during the year. Only when the year was over, the accounts had been added up and the actual profit of 196,000 rubles had been determined, did the administration finally respond: 'Here you are, your planned profit has been set at 196,000 rubles.' As a result the workers did not receive the bonuses promised.[4]

State farms, though shifted, in the main, to economic

[1] O. Pavlov, *Izvestiya*, 13.10.66, 5; *C.D.S.P.* 18 (41), 10–12.
[2] O. Pavlov, *Izvestiya*, 13.10.66, 5; *C.D.S.P.* 18 (41), 10–12.
[3] F. Tabeyev, *Izvestiya*, 18.11.66, 4; *C.D.S.P.* 18 (46), 27–28.
[4] V. Yefimov and K. Karpov, *Kommunist*, 15.10.66; *C.D.S.P.* 18 (51), 10.

accountability before the collective farms, continued to suffer from the restriction placed on their pattern of disposals, preventing them from entering into independent contracts. Appeals for greater freedom for state farms drew parallels with industrial enterprises, which could establish direct ties with consumers to dispose of above-plan output. 'Why cannot a state farm deliver its above-plan milk, vegetables or groats, say, directly to stores or workers' dining rooms?' it was asked.[1]

Economic accountability has greatly stimulated production on many collective farms, according to Soviet press reports. The farms that have made the most of the opportunities are those with the keenest management. Thus, one steppe farm produced sunflower seeds at about four rubles per centner but received 22·50 per centner. In addition to the basic and normal supplementary payments made at the end of the year, the farm operates a bonus scheme based on profits above the plan. A peasant can be deprived of these for good cause, such as absenteeism, petty theft, or bad work, which gives the management a powerful weapon. On this farm the peasants do not keep their own cows because the collective sells them milk more cheaply than they could produce it. A degree of specialization keeps costs low, but the collective avoids specialization on just a few enterprises in order to spread employment over as much of the year as possible for a large labour force.[2]

Not only the external financial relationships of the farms but also the internal operations have become subject to cost-accounting. The 23rd C.P.S.U. Congress recommended that collective farms should develop intrafarm accounting in all possible ways. In November, 1966, a farm at Livny, Orel oblast, reported its experiences.[3] It had established 'economic autonomy' for each of its twenty production units, with a purchase-and-sale basis for all transactions between the units. Adopting such a system necessarily brought difficulties among the workers who had been used to a casual way of working. 'The livestock raisers refused to sign an invoice for 2·5 tons of

[1] F. Tabeyev, *Izvestiya*, 18.11.66, 4; *C.D.S.P.* 18 (46), 27–28.
[2] G. Radov, *Literaturnaya gazeta*, 18.10.66, 2–3; *C.D.S.P.* 18 (47), 15.
[3] V. Koshelev, *Pravda*, 28.11.66, 2; *C.D.S.P.* 18 (48), 31–32.

fodder when only two tons had been delivered . . . the field workers argue with the repairmen over the high price of the latter's work, and so on.' But the managers claimed that they found a new sense of proprietorship among the peasants.

This farm also introduced new methods for forming the labour payment fund, which, for each subdivision, constituted the value of the annual output, less expenses at a predetermined proportional rate. Thus, increased gross profits would be reflected in higher pay increments backed by revenue.

Farms which have not changed over to full economic accountability have to allocate profits to various funds for reinvestment in the farm, insurance, payment of bonuses, etc. according to procedures laid down by the U.S.S.R. Council of Ministers in 1969.[1]

[1] *Izvestiya,* 12.1.69, 3; *C.D.S.P.* 21 (2), 23.

CHAPTER 4

Land and Climate

Only about 10 per cent of the land area of the U.S.S.R. is used for arable farming, about 2 per cent is cut for hay and a further 14 per cent. is classed as pasture, to which can be added the reindeer pastures of the north. The limitation of agriculture to about one-quarter of the total land area reflects physical conditions, especially the harsh climate of most of the Soviet Union. The relative importance of climate and soil in facilitating or restricting agriculture has been expressed by Russian scientists thus:

> Through the use of modern methods of climate evaluation, it has been determined that in an agricultural estimation of a territory, especially on a macroscale, climate is of decisive importance in the complex of natural factors. On a mesoscale, the importance of climate and soils is roughly the same, while on the microscale the decisive part belongs to soil.[1]

Although this statement needs qualification, it appears to be a valid generalization. Soil characteristics are themselves largely the result of climatic conditions, and another main soil-forming factor, vegetation, is also dependent on climate. The parent materials of soils may be formed irrespective of climate but their weathering is conditioned by climatic factors. On the macroscale, climatic effects are visible in the zonation of soils, but, locally, major differences in soil type reflect differing parent materials and relief, and exert important controls on agricultural practices.

Agriculture is affected by relief through altitude and slope of the land. The effects of altitude are felt mainly through climatic elements, while slope effects are also partly climatic in character, as in the creation of frost pockets or conditions

[1] Davitaya and Sapozhnikova (1969), 72.

of free air drainage. Slopes, however, directly affect agriculture through the gradients at which cultivation or animal grazing may be practised, and susceptibility to soil erosion, flooding and other hazards.

These effects are most pronounced in the mountain regions, especially in the Caucasus ranges and the many ranges of Central Asia, where the variety of relief forms is great and major changes of climate are associated with both altitude and aspect. Transhumance has been traditional in livestock rearing in these areas and is still practised on a limited scale. It no longer exhibits the features of true nomadism or the movement of whole families between different zones, but stock are moved between central farm settlements and distant pastures and are accompanied by herders. Alluvial fans and other lower slopes are commonly developed, sometimes with terracing and irrigation, for intensive cropping.

In the mountain regions of east Siberia and the Far East, agriculture is much more restricted in total by the severity of the climates and poverty of soils, but the further hindrances of altitude and steepness of slope are additional deterrents to development, and only the most favoured valleys offer opportunities for cultivation. Sunny slopes, however, commonly offer better conditions than valley floors which are marshy where permafrost prevents drainage and may also suffer flooding when the snow melts.

The Ural mountains are sufficiently high for altitude to result in moderately severe climatic conditions, and cropping is found only on the lower slopes. The southern Urals drive a wedge of uncultivable land between the steppes developed for grain and livestock rearing, so that only in a relatively narrow belt between Orenburg and Aktyubinsk is cultivation continuous from the European to the Siberian farmlands.

In the agricultural belts of the Russian plain there is little interruption of the farming by relief conditions, but there is substantial modification of the pattern in detail, resulting from varying glacial deposits in the north and centre. In some areas masses of boulders have had to be moved to permit cultivation, while the poor, sandy and rocky lands of morainic and outwash deposits are left to forestry or rough grazing. In the rich lands of the south, the loess is readily eroded and the rivers and streams have carved deep valleys and gullies which have greatly reduced the amount of cultivable land.

Alternating river valleys and interfluves create contrasting conditions for agriculture in a detailed mosaic of differing land types within the larger morphological units. Commonly the valleys themselves contain very varied soil and climatic conditions. The valleys of the Vyatka and Kama rivers and their tributaries in the Kirov and Perm oblasts and the Udmurt A.S.S.R. illustrate this. Between elevations of about 160 metres and 250 metres and with a moderately continental climate without marked local differences, except as caused by relief, the range of land types is considerable. Rocks represented include sandstones, shales, schists, clays and conglomerates, mostly of Permian age, covered by alluvial sands in places, but elsewhere exposed. The plant cover is mainly spruce and pine, with admixtures of birch and other deciduous trees and shrubs. The interfluves are generally inhospitable, but some agriculture is possible where a thin covering of sands yields sandy-loamy soils. High terraces along the major rivers retain soils with thick humus horizons, relicts of earlier floodplain conditions. Much of this land is under the plough, but subject to erosion hazards. Tributary streams have cut through these terraces and left their own terraces with sandy soils suitable only for forest growth. Marshy soils, some capable of development for pasture, occur in the valley bottoms. There are also old alluvial sandy plains with a dune type of surface configuration mostly under forest and scrub. Thus, agricultural land is very limited, restricted to suitable morphological types, on which favourable soils have developed. This great local variety, whose major effect is to reduce the area available in practice for farming because of unsuitable terrain, must be remembered when interpreting the broad regional scene in terms of climate and soil.

THE CLIMATIC ENVIRONMENT FOR AGRICULTURE

In most parts of the U.S.S.R., agriculture is handicapped by unfavourable climate. The great size of the country results in continental conditions, with air masses and winds generally outward-moving from high-pressure areas in winter, while inward-moving summer winds lose most of their moisture before penetrating far inland. The northerly position of most of the country results in generally harsh temperatures during much of the year. Variations in weather, particularly of precipitation, make for great fluctuations in yields from year to year, though day-to-day variation is much less marked than in areas of more maritime climate.

For cultivation, the intense winter cold experienced in Siberia and the only slightly more moderate conditions normal in European Russia are less important than the shortness of the summers, lateness of spring and prevalence of water shortages. Animal husbandry is also handicapped by water shortages, extremes of temperature and the long periods of cold conditions which necessitate expensive shelters.

Such generalizations must, of course, be modified in varying degree for different regions. The Baltic republics and adjacent parts of the R.S.F.S.R. and Belorussia are relatively well endowed with moisture, reflected in the importance of dairying in their agriculture, but inadequate drainage limits the areas that are agriculturally developed. Southerly areas from the Ukraine to the Far East enjoy high summer temperatures but in few areas does summer precipitation offset high evaporation and transpiration, so that irrigation is necessary for intensive farming. Only in the so-called sub-tropical areas of Transcaucasia and in the Maritime kray in the Far East do high summer temperatures coincide with ample rainfall. Most areas have summer maxima of precipitation but only rarely is the total amount satisfactory for agriculture.

In considering the effects of climate on agriculture, we are studying, above all, the effects of solar radiation and precipitation. It is upon these two inputs and the corresponding outputs from the earth's surface that such secondary conditions as soil temperatures and soil water depend.

Organic matter is produced in plants by photosynthesis, utilizing radiation energy, and by chemosynthesis, utilizing chemical energy. Photosynthesis is essential in producing the biological mass, while chemosynthesis is important mainly in nitrogen transformation and other processes. In photosynthesis plants use water, atmospheric carbon dioxide, and small amounts of minerals from the soil for building organic matter. In plant growth, only very small fractions of the available resources of energy and water are utilized, but there is a close correlation between different zones of natural vegetation and the distribution of radiation balance and precipitation.

The agricultural significance of the climatic variations over the area of the Soviet Union may thus be examined primarily through the concepts of the heat balance and water balance, with subsidiary consideration of other elements, notably wind.

Budyko, in his study of the water and heat balances, observed that in climates with a limited growing season, vegetation grows only during the period with positive radiation balance, minus that portion of this period in which the positive radiation balance compensates the negative radiation balance of the cold months, and, furthermore, that negative annual sums of radiation are found only in high latitudes, where vegetation is extremely limited. From these conclusions, and the significance of the radiation energy balance at the level of the active surface for transpiration and photosynthesis, he argued that the radiation balance of the underlying surface should be regarded as the 'base of energy' for the production of natural vegetation.[1] There is, too, an optimum of soil moisture, below which the plant does not develop sufficient assimilating surfaces and above which oxygen starvation inhibits root development.

Budyko, therefore, agreed with Grigoryev that the productiveness of natural vegetation reaches a relative maximum at some optimal interrelationship of the energy base and precipitation, and decreases as the interrelationship changes in either direction from the optimum.[2]

The combination of heat and water balance data forms the basis of current Soviet maps of climatic regions and the climatic resources available for plant growth.[3] In both cases, as presented in the *Physical-geographical atlas of the world*[4] and the *Atlas of agriculture of the U.S.S.R.*[5], respectively, available warmth is shown by accumulated temperatures,[6] the threshold temperature for these calculations is 10 degrees C. (50 degrees

[1] Budyko (1956) 1958, 186. [2] Budyko (1956) 1958, 189.

[3] For a brief description, see Grigoryev and Budyko (1960).

[4] *Fiziko-geograficheskiy atlas mira,* (1964), 203.

[5] *Atlas sel'skogo khozyaystva SSSR* (1960), 46–47.

[6] Owing to the lack of measurements of actual net radiation and other heat balance components, Budyko used measured parameters such as temperatures and sunshine duration to approximate the required components. A similar approach has been used here in delimiting climatic regions and zones.

F.). There is, however, a difference in the data used in these two maps, those of the climatic regions being based on the sum of temperatures at the earth's surface during the period when the air temperature remains above 10 degrees C. The accumulated temperatures shown, therefore, are generally higher than in the other map, which uses accumulated air temperatures.

The work of Budyko and others has been extended into the science of agroclimatology, in which the practical application of climatological knowledge to evaluating zonal potentials for agriculture and the climatic requirements for different crops and enterprises is emphasized. These studies have helped to reveal the conditions needed for long-term successful cultivation and the possibilities of introducing desired crops into other, more marginal, regions.[1]

Warmth, moisture and light are the principal components of the climatic resources available for agriculture, other climatic features, such as wind and cloudiness, being of subsidiary importance, weakening or reinforcing the effect of the fundamental components. These latter factors may assume dominant importance locally, when they become extreme, as do winds which cause or exacerbate drought and the removal of snow from the soil when its cover is needed, but warmth and moisture provide the basic indices used in agroclimatology and the delimitation of agricultural regions.

The importance of accumulated temperatures in the growing season is shown in Table 5, which gives the requirements of selected varieties of various crops.

Table 5 shows that the sum of temperatures required varies considerably according to the variety of crop and the objective for which the crop is being raised. The lower requirements for eastern Siberia and the Far East are explained by smaller fluctuations permitting finer limits and by the clearer continental conditions. In some cases, as in the sub-tropical crops, distribution is conditioned less by accumulated temperatures than by winter conditions. In general, the temperatures shown should ensure satisfactory results in nine out of ten years.

[1] For methodology, see Shashko, in *Soil-geographical zoning of the U.S.S.R.*, Academy of Sciences of the U.S.S.R. (1962) 1963, 378–445. For a brief account, see Davitaya and Sapozhnikova (1969).

In addition to accumulated temperatures, the availability of warmth at critical periods of plant growth, the risk of damage from heavy frosts and the number of days available for growth are important aspects of temperature. Some of these limits,

TABLE 5

CROP REQUIREMENTS IN ACCUMULATED TEMPERATURES (Above 10°C.) FOR RIPENING IN COMMERCIAL CONDITIONS

Crops	*European U.S.S.R. and western Siberia (day-degrees C.)*	*Eastern Siberia and Far East (day-degrees C.)*
Vegetable crops on sheltered ground	400	400
Turnips, cabbage (e)	800	700
Barley (e), winter rye (e), in warmer locations	1,000	800
Barley (e), peas (e), flax for fibre	1,200	1,000
Oats (e), barley (m)	1,400	1,200
Spring wheat (e), winter wheat, maize (m) for green feed, sugar beet for fodder	1,600	1,400
Barley (l), oats (l), maize (l) for silage, millet (e), wheat (m)	1,800	1,600
Spring wheat (l), sunflower (e) for seed, sugar beet for sugar	2,000	1,800
Maize for grain (e), beans (e), millet (l)	2,200	2,000
Beans (l), sunflower (l) for seed, apricots	2,800	2,400
Maize (m–l) for grain, rice (m), grapes	3,200	—
Soya beans (l), ground nuts (e), sorghum (l), figs	3,600	—
Cotton (e), lemons, tangerines	4,000	—
Cotton (m), rice (l), grapes (l)	4,400	—
Olives, oranges, jute	4,800	—

e—early, m—medium, l—late ripening varieties.

Sources: Shashko (1962) 1963; Davitaya and Sapozhnikova (1969).

together with the main seasons at which the crops concerned require ample supplies of water, are shown in Table 6. This table also shows a range for accumulated temperature requirements, which even when allowance is made for differing requirements of early and late maturing varieties, emphasizes that the concept of accumulated temperatures does not give an exact guide to the needs of particular plants. Much depends

TABLE 6

CLIMATIC REQUIREMENTS OF SELECTED CROPS

	Accumulated temperatures needed in the vegetative period. (day-degrees C.)	*Number of days required for the vegetative period*	*Temperature below which plants are likely to be damaged (degrees C.)*	*Seasons of maximum demands on soil water*
Spring barley	950–1,450	60–100	−6	Spring, early summer
Winter rye	1,000–1,250	80–120	−20	Spring, early summer
Oats	1,000–1,600	80–120	−6	Spring, early summer
Peas	1,050–1,550	60–120	−6	Spring, early summer
Winter wheat	1,150–1,500	80–120	−10	Spring, early summer
Buckwheat	1,200–1,400	60–100	−1	Summer
Spring wheat	1,200–1,700	80–120	−6	Spring, early summer
Flax for fibre	1,200–1,600	80–100	−4	Spring, early summer
Potatoes	1,200–1,800	60–120	−2	Summer
Lucerne	1,000–2,000	100–120	−15	Summer
Flax for oil	1,200–2,000	80–120	−4	Spring, early summer
Red clover	1,200–2,000	80–120	−10	Summer
Timothy	1,200–2,000	80–120	−20	Summer
Hemp (Central Russian)	1,300–1,800	100–120	−4	Summer
Millet	1,400–1,950	60–120	−2	Summer
Sunflower	1,600–2,300	100–140	−4	Summer
Maize (milk-wax stage)	1,800–2,400	80–120	−2	Summer
Hemp (Southern)	2,000–2,800	120–160	−4	Summer
Sugar beet	2,000–2,800	120–160	−4	Summer
Maize, for grain	2,100–2,900	120–160	−2	Summer
Sorghum, for grain	2,250–2,850	120–160	−2	Summer
Soya	2,450–2,950	100–160	−1	Summer
Rice	2,200–3,200	100–180	−1	Summer
Groundnuts	2,500–3,500	120–180	−1	Summer
Cotton	2,900–4,000	140–180	−1	Spring to autumn

Sources: Stepanov (1957); Sapozhnikova and Shashko (1962).

on the combination of climatic elements in any given year, or of climatic and edaphic and other conditions.

CLIMATIC REGIONS

The following examination of the relationships between climate and agriculture in the Soviet Union is based on the regions and zones established by the water balance and accumulated temperatures (Figures 3 and 4, Table 7). For precision the following terminology will be used:

CLIMATIC REGION (or, in this chapter, sometimes simply '*region*')—a region as outlined and numbered in the map of climatic regions, referred to above and reproduced in Figure 4.

MOISTURE AREA—a group of regions having the same initial numeral, (from section (1) in Table 7), indicating the positive or negative state of the annual moisture balance.

HEAT ZONE—an area between stipulated isolines of accumulated air temperatures as shown in the plate of the *Physical-Geographical Atlas of the World* facing that of climatic regions and in the map of agroclimatic resources in the *Atlas of Agriculture of the U.S.S.R.* (Figures 3 and 4) the index figure used being as in section 2 of Table 7.

SUB-REGION—a division of the climatic region distinguished by one of the suffix letters used to indicate the nature of the winter experienced (from section (3) of Table 7).

TABLE 7

CLIMATIC REGIONS OF THE U.S.S.R.

(After Budyko, *Fiziko-geograficheskiy atlas mira*, Moscow, 1964)

(1) *MOISTURE CHARACTERISTICS*

	*Aridity index**	*Vegetation type*
I Excess moisture	Less than 0·45	Arctic desert, tundra, wooded tundra, alpine meadow
II Humid	0·45–1·00	Forest.
III Inadequate moisture	1·00–3·00	Forest-steppe, steppe, semi-desert.
IV Arid	More than 3·00	Desert.

* The aridity index is the ratio of evaporability (related to radiation balance, temperature and humidity) to precipitation.

TABLE 7—*continued*

(2) *WARM SEASON TEMPERATURE CONDITIONS*

	Sum of temperatures of earth's surface during the period in which the air temperature is above 10°C. (day-degrees C.)*	*Vegetation type*
1 Very cold	Air temperature below 10°C. all year	Arctic desert.
2 Cold	Less than 1,000	Tundra, wooded tundra.
3 Moderately warm	1,000–2,200	Coniferous forest, alpine meadow, mountain steppe and steppe in Siberia, mountain deserts of the Pamirs.
4 Warm	2,200–4,400	Mixed and broadleaved forest, forest-steppe, steppe, northern desert.
5 Very warm	More than 4,400	Sub-tropical, desert.

* The sum of the temperatures at the earth's surface is, as a rule, greater than the sum of air temperatures in the same period.

(3) *WINTER CHARACTERISTICS*

	Mean January temperature in degrees C.	*Maximum 10-day snow cover*
A Severe, little snow	below −32	Less than 50 cm.
B Severe, snowy	below −32	More than 50 cm.
C Moderately severe, little snow	−13 to −32	Less than 50 cm.
D Moderately severe, snowy	−13 to −32	More than 50 cm.
E Moderately mild	0 to −13	
F Mild	above 0	

Moisture area I includes Arctic desert and tundra lands. Region I–1 has an Arctic climate (excess moisture and very cold) and no agricultural value, and region I–2 (excessively moist and cold) is of negligible significance agriculturally but is used extensively for reindeer herding and hunting.

Very hardy crops and cattle are raised in favourable places in this region, but accumulated air temperatures do not

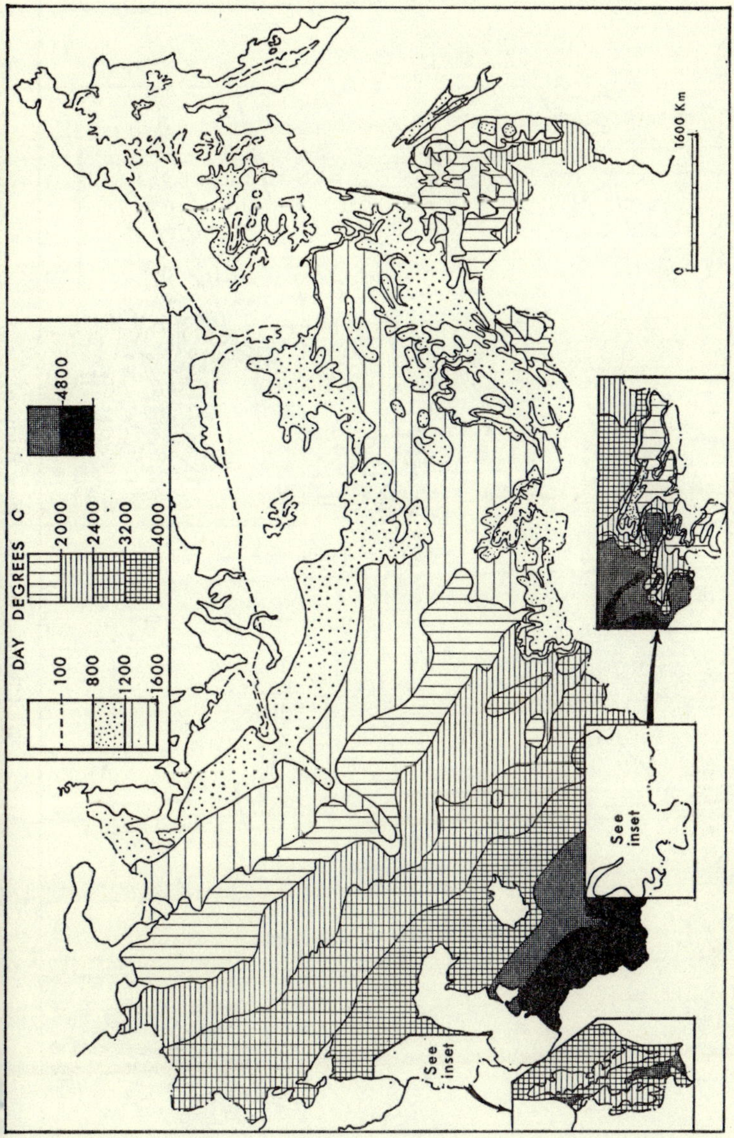

3. Accumulated temperatures (above 10°C.)
Source: Fiziko-geograficheskiy atlas mira.

generally exceed 1,000 day-degrees. This is especially true of the valleys of the northern margins of the tayga and of the

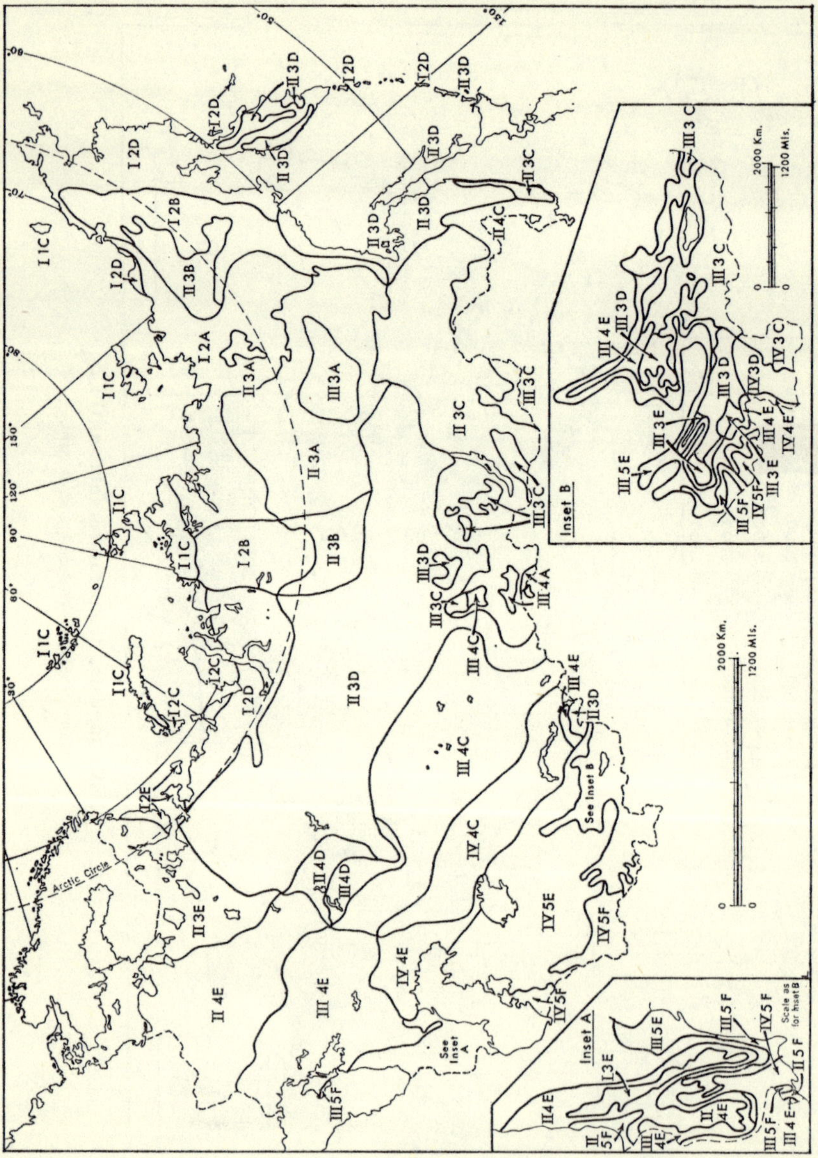

4. Climatic regions (after Budyko)

mountain regions of eastern Siberia (I–2A) and the Far East (I–2B), and the coasts of Kamchatka (I–2D).

Moisture area II covers the main belt of the tayga and the broadleaved forest areas, so considerable differences occur within it. None of it is classed as cold or very cold, these terms being reserved for the more extreme northern areas, but region II–3 records accumulated temperatures of no more than 800 day-degrees in its northern parts. Sub-regions II–3A and II–3B in central and eastern Siberia include territory receiving up to about 1,200 day-degrees but differ from region I–2 only in that pockets of meadow, cultivation and livestock rearing are more common, with reindeer herding and hunting still the dominant rural occupations. Sub-region II–3C (in eastern Siberia and a small part of Maritime kray) includes much mountain country, and accumulated temperatures may be as low as 800 day-degrees but as high as 1,800 or even more in the valleys. Sheep rearing and cattle rearing, with beef or beef-dairy types predominating, occur in these more southerly valleys. Sub-region II–3D, with a little less severe climate though snowier winters than the preceding sub-regions, extends from central Siberia across the Ural mountains to Arkhangelsk and the headwaters of the Volga, with a detached area in the Far East, including parts of the Pacific coast, the eastern part of Maritime kray and Sakhalin. In this climatic sub-region, the agricultural pattern changes gradually at about the 60th parallel west of the Ural mountains and somewhat further south in Siberia, with the change to a sum of temperatures exceeding 1,600 day-degrees above the 10-degree threshold. In the colder zone, corresponding to the main belt of the tayga, agriculture is virtually limited to stock raising on the river-valley meadows, with supplementary fodder crops. In the more favoured areas, however, notably the Northern Dvina and neighbouring valleys, with the frost-free period exceeding 90 days, sowings include rye, oats, barley, wheat, potatoes, beans and other vegetables. Wheat may occupy 5–10 per cent. of the sown area, but only spring wheats are practicable, yielding at best about 10 centners per hectare. In the European area of this zone the sown area decreased between 1940 and 1966, despite the continued improvements in fast-growing strains of grains and grass, but west Siberia recorded a significant increase in its modest proportions under the plough.

The south-western part of climatic sub-region II–3D, and the southern part of sub-region II–3E, where nearness to the Baltic Sea moderates winters, record over 1,600 day-degrees. The isoline of this sum of temperatures corresponds roughly with the transition to more complex forms of agriculture. All these agricultural regions continue southward in the warmer climatic sub-regions II–4D and II–4E, to about the isoline of 2,000 day-degrees. The greater summer warmth of this area is attained without a very high number of hours of bright sunshine and the light is generally diffused, conditions which promote vegetative growth without early flowering and fruiting. Hence, plants whose stem and leaf products are valuable are encouraged. Flax for fibre is an important cash crop, and silage and hay crops come into this category. Numerous vegetable crops for human consumption, makhorka, mustard and other minor crops, increase the variety of this zone. Potatoes and root crops which develop slowly through the season utilize well the even distribution of precipitation and are major crops throughout sub-region II–4 E and the southern zones of II–3 D and II–3 E.

Sub-region II–4 E is mild enough, given the moderate snow cover, for winter wheat as well as spring wheat. In the zone of up to 2,200 day-degrees winter wheat yields are, however, only about one-half those of the warmer Ukrainian areas, being at best about 12 centners per hectare. Good spring wheat crops yield about 10 centners per hectare. Rye consequently remains important in all except the most southerly part of the sub-region. In the 1,600–2,200 day-degree zone, however, wheat is the grain crop showing the greatest increase in utilization, its area increasing in most oblasts by between 10 and 40 per cent. between 1958 and 1965. Maize is grown to be harvested green, but oats and barley are preferred as fodder crops. With more than 1,800 day-degrees, buckwheats are useful and are grown in the southern part of sub-region II–4 E and in the rather more continental conditions of II–4 D, where winter wheat is less reliable.

In these sub-regions, annual average precipitation exceeds

400 mm. and more than half falls in the growing season, while the frost-free period is typically 120–150 days. Perennial grass and clovers give high yields in these well-watered areas of low evaporation and occupy more than 20 per cent. of the sown area in many oblasts. In the drier zone east of Moscow lucerne is favoured. The improvement of the fodder base with silage, green crops and sown pastures has permitted substantial increases in production of milk, beef and pork. Production increases have been demonstrated particularly in the Moscow area, which is in sub-region II–4 E, but in other parts of this non-chernozem zone an undue proportion of the land remains in unimproved hay meadows and poorly developed cropping.

In Siberia, sub-region II–3 D includes the aspen-birch fringe of the forest zone, where livestock rearing, especially of dairy cattle, and spring wheat growing are prominent. The frost-free period is only 100–135 days and accumulated temperatures are 1,600–1,800 day-degrees. Precipitation here falls to 250 mm. per annum, so that although two-thirds or even three-quarters falls in the growing season, lucerne is preferred to clover and millets are grown as far north as Tomsk. The industrial regions of the Novosibirsk and Kuznetsk areas have stimulated production, especially of milk, meat and vegetables.

Cooler summers prevail in the mountain areas of central and eastern Siberia and agriculture both west and east of Lake Baykal is devoted mostly to livestock rearing, particularly cattle for beef and sheep for wool. In smaller areas of cultivation, with accumulated temperatures locally over 2,000 day-degrees, grains and roots are important. Some of these areas are classed climatically as III–3 C. The Amur valley from Blagoveshchensk to Khabarovsk and the Ussuri valley are climatically II–4 C. Maize, oats, soya beans, rice and other crops are practicable in this humid and relatively mild area, but excess summer rain frequently damages grain crops. The lower Amur valley and coastal areas of Maritime kray northwards are in climatic sub-region II–3 D and less developed agriculturally.

Moisture area III is characterized by inadequate precipitation and the natural vegetation varies from wooded steppe to steppe and semi-desert. Accumulated temperatures exceed 2,000 day-degrees in the European areas and 1,800 day-degrees in western Siberia, rising to 3,600 day-degrees in the north Caucasus area.

Water conservation as well as irrigation is very necessary. Much of the summer precipitation falls in heavy downpours, which may reach an intensity of 4 mm. a minute. Maxima recorded in 24 hours are 186 mm. at Sochi and 104 mm. at Kiev, whereas at Leningrad not more than 60 mm. has been recorded.[1] Such heavy rain is largely lost by rapid run-off in the dissected relief conditions of the area.

Region III–4 and the more humid southern zone of II–4 E, where accumulated temperatures also exceed 2,400 day–degrees, covers the main grain and sugar beet areas of the Ukraine and adjacent parts of Belorussia and the R.S.F.S.R. central chernozem zone. Warmth-demanding crops such as tobacco, sunflower, melons, grapes and other fruits grow well, provided water supplies are assured, with irrigation where necessary. Much of the area has over 70 per cent. of the total farm land under the plough. Most of the great cultivated belt stretching from Moldavia to the upper Ob is included within this region.

Winter wheat increases from about 10 per cent. of the sown area on the northern margins of sub-region III–4 E to over 40 per cent. on the Black Sea steppes and in the Stavropol area, declining sharply in Stavropol kray as annual precipitation falls to some 350 mm., with evaporation double this figure. The spring wheat belt begins in the north of this sub-region and continues through III–4 D, north of Kuybyshev, and eastwards through III–4 C, the grainlands of west Siberia and north Kazakhstan. This belt also contains nearly all the millet areas. It has been noted that millet suffers from drought 10–15 per cent. less frequently than spring wheat, a coefficient of available moisture of 0·6 for spring wheat and 0·4 for millet indicating the drought level for these crops, below which harvest yields decline.[2] Only in the southerly and humid or irrigated regions are conditions suitable for full ripening of maize, and sub-region III–4 E

[1] Borisov (1959) 1965, 231.
[2] Borisov (1959) 1965, 174.

grows most of the region's maize, with decreasing proportions east of the Don and in Siberia (sub-region III–4 C).

Oats and barley also remain important in the fodder base of these two sub-regions and III–4 D. There is close correspondence between the boundaries of moisture area III and the replacement of fibre by seed varieties of flax, and almost all the sunflower growing is within it, except for the Central Asian and Far East areas. These regions, then, are important producers of the major oil seeds, while castor oil and other minor oil seeds also occur mainly within their boundaries.

Though supporting fewer livestock per 100 hectares of arable land than sub-region II–4 E, region III–4 also is important for stock rearing. In the better-watered sub-region III–4 E one cow or steer per 2 hectares of sown land is not uncommon. As might be expected, cows are proportionately fewer in this drier area than to the north, but milk yields compare favourably with the northern areas. The eastern half of this sub-region is one of the most important for sheep rearing in the Soviet Union. The dry conditions of Stavropol kray are well suited to the Merino and its crosses, and fine wool is the main objective of sheep rearing here. Pigs also are less numerous than in sub-region II–4 E but very important still in the western areas of sub-region III–4 E and near the larger cities in the eastern areas.

Moisture area IV is the vast desert and semi-desert area, with a dryness index of more than 3·00, which extends from the eastern boundary of Stavropol kray and Volgograd oblast across north Kazakhstan to the mountains of Central Asia. Throughout this zone the average July temperature exceeds 24 degrees C., the growing season is 200 to 290 days and the frost-free period 150 to 250 days. Potential evaporation rises from 800 mm. in the north to over 1,400 mm. south of the Aral Sea, which is from four to over fifteen times the average annual precipitation. Hence, the value of the high solar radiation of 120 to over 160 kilolangleys can be realized only where oases and artificial irrigation provide ample water supplies. Rearing cattle and sheep by extensive methods remains the only agricultural enterprise over wide areas. Pasture area per head of cattle averages over 10 hectares in most of this vast region. Sheep are reared widely, with

Karakul fleeces the principal objective in the very hot and dry areas of Turkmenistan and Uzbekistan.

With assured water supply, as in the Syr Darya valley, cotton, wheat, maize, oats, barley, rice, sugar beet, oil seeds and other crops, including numerous vegetables and fruits, are grown. Grass, lucerne and fodder crops are grown for livestock. These irrigated areas are especially important for cotton, which needs at least 180 days above 10 degrees C. (mean daily temperature) and accumulated temperatures above this level of 3,000 day-degrees.

Flanking moisture area IV east and west is the complex of mountain and valley climates of the Caucasian and Central Asian ranges. Higher relief induces higher precipitation and lower temperatures and evaporation, so that these areas are classified as separated parts of moisture areas II and III, while their southerly location makes possible a warmth rating of 3 or 4 for the lower and intermediate altitudes. Crops and livestock of temperate zones, therefore, are appropriate and many of the valleys and slopes are highly productive. High precipitation, summer warmth and mild winters in the Kolkhid lowlands lead to a classification of II–5 F, and the conditions particularly favour the Mediterranean and sub-tropical fruits and crops, including grapes, citrus fruits and tea, for which it is renowned. Severe frosts do occur, however, in winter, necessitating precautionary measures in the orchards. The smaller area of the Lenkoran lowland is similar and is classified II–5 F, but most of the Kura valley lowland is classified IV–5 F, with cotton, grains and livestock the main agricultural products. Intermediate moisture conditions result in a classification of III–5 E and III–5 F on higher slopes in the Kura valley and northwards along the Caspian shore, and fruit, vegetables, maize and other grains are important. These crops appear also in the broad belt of rising land north of the Caucasus, where, however, the emphasis is on livestock rearing, from dairy and beef cattle on the lower land near Pyatigorsk and Ordzhonikidze to sheep rearing on the higher pastures. Somewhat similar conditions occur in Armenia, the central area of which is classed as II–4 E and numerous crops and livestock feature in the agricultural system.

The mountains of Central Asia provide regional belts of

higher humidity than the desert areas, their relief producing the release of moisture from air masses which have crossed the desert without precipitation occurring. The altitudinal zoning is fairly regular. In the foothills, conditions for cultivation are exceptionally good. The land between about 500 and 1,000 metres, which includes such important towns as Alma Ata, Frunze and Dushanbe, is classed as III–4 E and III–5 E, with accumulated temperatures of about 2,400 day-degrees. Namangan in the upper Fergana valley enjoys the longest frost-free period in the Central Asian valleys, 236 days, from its southerly exposure and protection from the north by the mountains.[1] The interval between the first spring and last autumn days with an average daily temperature above 14 degrees C., which is favourable for cotton growing, amounts here to about six months. Dushanbe also has cotton growing but most of this area is outside the climatic limits for cotton, and grains, sugar beet, oil seeds, fruit and livestock provide the main enterprises. Temperatures and frost-free periods decrease rapidly with altitude, and the cooler III–3 C sub-region is dominated by livestock rearing, with fine-wool and mutton production, horse breeding, and cattle of beef rather than dairy characteristics.

The result of the interaction of the heat and water balances over most of the Soviet lands capable of growing crops is a deficiency of moisture. Evaporation is greatly increased by winds, and consequently droughts, and winds which exacerbate them, are major physical opponents of farming, so that agro-climatologists have given much attention to them. Snow cover, although a hindrance to cropping and grazing, is more an ally because it provides valuable moisture and also limits the freezing of soils. Frost, however, is a serious problem in most of the Soviet Union and has accordingly been studied widely.

DROUGHT

In 1913, Rotmistrov, an agricultural research director of the Odessa Agricultural Experimental Field, defined drought as a sufficiently long rainless period for soil moisture to be totally

[1] Borisov (1959) 1965, 212.

exhausted by plant assimilation. This classical definition has been modified, but it indicates the essential situation—a lack of precipitation resulting in insufficient soil moisture and retarded plant growth. Using as an index of humidity the ratio of total rainfall to the sum of temperatures multiplied by 0·1, a value of 0·6–0·7 is indicative of drought, with a corresponding decrease in harvest yields of 20–25 per cent.[1]

In European Russia two pronounced centres of droughts are distinguished, one in the south-east, the other in the Ukraine. Droughts generally occur simultaneously in these areas, the boundary of the drought-afflicted area passing from Zhitomir through Moscow and Kostroma to the Urals near Perm. The frequency of droughts increases towards the south and south-east, reaching a maximum in European territory between Volgograd and Uralsk and the Caspian coast.[2] One in every three or four years is a drought year in the steppes, according to Borisov.[3] A similar frequency of droughts obtains in the southern districts of the central Ural mountains, and severe droughts occur there every 8–12 years.[4] Droughts are particularly likely when the preceding winter has brought little snow and has given way to an early and dry spring, so that moisture has not adequately accumulated in the soil.

To combat drought the extension of forest belts has been urged, on the grounds that they increase precipitation and conserve snow, benefiting soil moisture reserves. Irrigation is, however, more reliable for ensuring steady harvests. According to Babushkin, the hazard of soil droughts is practically non-existent in the irrigated cotton growing area of Uzbekistan, but atmospheric drought continues to create a hazard.[5] Atmospheric droughts are arbitrarily defined by the vapour pressure deficit, and have occurred in 20 out of 25 years in some districts of the Bukhara and Tashkent areas. With wind, the atmospheric drought becomes a *sukhovey*, which is discussed in the following section.

[1] Alpatev (1950), summarized in Selyaninov (1957) 1963, Borisov (1959) discusses other measures of drought and also places emphasis on the agroclimatic characteristics of drought.

[2] Alpatev (1950), Selyaninov (1957) 1963.

[3] Borisov (1959) 1965, 167.

[4] Borisov (1959) 1965, 193.

[5] Babushkin (1957) 1963.

THE WIND HAZARD

Unlike precipitation and sunshine, wind is almost entirely adverse to farming, though, of course, it is an essential part of circulation and moisture distribution. Wind velocity varies greatly in the U.S.S.R., averaging from 1·0 m/sec. in Siberia to 10·5 m/sec. in the vicinity of Novorossiysk. Except in east Siberia, where winters are calm, wind velocities are highest in winter, with summer the calmest season, so that grain growing is not greatly hindered by winds. The semi-arid areas, however, experience strong winds, 4·0–4·5 m/sec. or more, in the steppes and semi-deserts, and 3·5–4·0 m/sec. in the forest-steppe, compared with 3·0–3·5 m/sec. in the forest zone. In general, however, averages have little significance for agriculture, although persistent high winds can hinder grain growing and be harmful to livestock. More commonly, it is abnormal or local winds which are most damaging to farming.

Most prevalent and harmful because they cause rapid wilting of plants are the warm, dry winds known collectively by the name of 'sukhovey' (lit. 'dry wind'). Sukhovey conditions have been defined variously, for example, as when relative humidity of the air is below 30 per cent. with the temperature above 25 degrees C. and a wind velocity of over 5 m/sec. at the wind-vane level. The occurence of similar effects with different humidities and temperatures has encouraged attempts to define the conditions more generally, as, for example, that a sukhovey is indicated when the vapour pressure deficit of the air reaches or exceeds 17 mm. with a wind velocity of 5 m/sec. or more.[1] This and other definitions based on vapour pressure deficit interpret the sukhovey essentially as a special case of drought. The difference between the vapour pressure deficit at plant height (in the stand) and at a height of 2 m., where it is normally greater, is reversed when sukhovey damage occurs, and, if this value steadily exceeds 2 mm., considerable damage occurs except at the stage of complete ripeness.[2] There is no entirely satisfactory definition of a sukhovey, but the mean 24-hour value of the

[1] Kulik (1957) 1963, 59–64.

[2] Tsuberbiller (1957) 1963, 73. Tsuberbiller also indicates the usefulness of the more exact but more complex evaporimetric coefficient as proposed by Skvortsov (1949). A steady decrease of this coefficient to 0·3 (when the evaporability is three times as high as actual evaporation) leads to a decreased crop yield.

vapour pressure deficit appears to provide the material for the best approximation when taken in conjunction with the duration of the wind and drought conditions.[1]

The sukhovey is a feature particularly of the steppes and wooded steppes of the European areas, the west Siberian plains, the Kazakh-Turan lowlands and elsewhere in Soviet Central Asia and east Siberia. The sukhoveys occur typically on the southern, south-western and western peripheries of anticyclones, the air masses coming from the west, north and north-east and not from Central Asia as was formerly supposed.[2] There are few obstacles to these winds on the smooth, open steppes and their effects reach as far as Moldavia. On the Volga steppes the average number of days with sukhoveys has been reported as: April, 0·5; May, 1·2; June 0·9; July, 0·6; August, 1·1; September, 0·4; October, 0·1.[3] This relatively low frequency may indicate a conservative definition of the phenomenon in this case. The Crimean steppes, it is reported, suffer sukhoveys on an average of 44 days in the warm period, with wind velocity of more than 10 m/sec. on seven such days.[4]

In west Siberia and Kazakhstan, sukhoveys occur most frequently in May and June. The number of days with sukhoveys varies considerably with location. Thus, in the steppes of northern Kazakhstan, between April and September, 3–5 is normal in the north, rising to 12–13 in the south around Karaganda and Osakarovka. In north Kazakhstan sukhoveys generally last only one day but in western Kazakhstan often 4–6 days. Sukhoveys and droughts are blamed for the absence of an exportable surplus of crops from the drier region further south, where livestock rearing is the main enterprise, there being as many as 17–19 days of sukhoveys in the warm season in the west of this area. The sukhovey frequency rises to 20–40 days east of the Caspian Sea, to 60 days in the Golodnaya steppe, with up to 100 days in particularly droughty years. Varieties of sukhoveys which have been discerned in Kazakhstan include cold sukhoveys in early spring, which do not directly effect plants but deplete soil moisture reserves, and föhn sukhoveys. The latter winds, hot and dry after passage over

[1] Feldman (1957) 1963, 98.
[2] Evseev (1957) 1963.
[3] Borisov (1959) 1965, 171.
[4] Borisov (1959) 1965, 231.

mountains, are particularly damaging to orchards in blossom and to cereals when flowering or forming grain.[1]

All crops may suffer from the effects of sukhoveys.[2] Fruit and vines are particularly vulnerable, and damage to or even destruction of crops is frequently reported from Kazakhstan, Central Asia and the Caucasian sub-tropical areas. Pastures and hayfields also suffer damage. The importance of protecting agricultural lands by shelterbelts and of developing irrigation to combat these hot, dry winds and associated droughts is fully appreciated by Soviet agricultural scientists, but the work is still in its infancy.

Conditions akin to those of the sukhovey type of wind occur in Central Asia also with the 'harmsil' or 'afganets', which are high winds from the mountain areas, bringing heat and dust in summer and snowfalls and blizzards in winter. In southern Tadzhikistan the afghanets may reach a speed of 15 m/sec. Winds with föhn characteristics occur frequently in the Caucasus areas, the Crimea and about Lake Baykal and cause crop damage in summer and snow melt in winter, sometimes with exposure of the soil to heavy frosts. Dry, dusty winds in the Baku district may blow for as long as nine or ten days without interruption.

By contrast, cold winds bring freezing conditions to the sub-tropical lowlands in winter, and in the Caucasian grape growing areas the vines may have to be earthed over to protect them. The 'bora' at Novorossiysk contributes to the Soviet record of high average of winds recorded there, while the winds known as the 'burana' bring exceptional blizzards to the steppes and tundras. Because of the severity of such winter winds frameworks are built between buildings on farms and covered over in the winter to offer protection to workers and livestock even in such southerly areas as the north Caucasus plains.

SNOW

For agriculture, snow has both negative and positive value, negative in inhibiting cultivation and livestock rearing,

[1] Samokhvalov (1957) 1963, 49–50.

[2] Lydolph (1964) gives a table of vapour pressure deficits at which differing degrees of damage are recorded for maize, sunflowers, potatoes and grain based on Tsuberbiller's reports.

positive in bringing moisture to the land, and protecting winter-sown crops. A deep snow cover limits soil freezing, and the earlier soil thaw that results enables spring cultivation to be started early.

Snow cover lasts at least five months in the northern agricultural regions and in most of Siberia, and three to four months in the steppes west of the Ural mountains and in the Baltic republics. The depth reached in this last area is 20–40 cm., with 40–60 cm. in the European mixed-forest zone and in the Siberian steppes from Sverdlovsk to Krasnoyarsk. Only in Central Asia does the period of snow cover fall to less than a month and there only in the desert areas east of the Caspian Sea.

The importance of snow depth in regions of heavy frost lies in the protection snow gives against freezing. In the chernozem belt of Kazakhstan, for example, the snow cover varies from about 16–17 cm. in the south and south-west to 30–35 cm. in the north and north-east. On the higher ridges and interfluves the soils freeze to a depth of as much as 120–180 cm., on lower ground to 50–100 cm. The soils do not completely thaw out until early summer, even as late as early July.[1] Added to the liability of the area to droughts, this imposes severe limitations on the practicable intensity of land use and increases the need for fallowing.

In easterly parts of the forest-steppe areas of Krasnoyarsk kray, the snow cover is only 20–25 cm. deep and on the open steppe freezing takes place down to 3·5 m. and thawing is completed only in July or even August.[2] Freezing affects soil development and increases gleying processes, which become pronounced even in areas of low precipitation and good drainage, as in the grey forest soils of the Altay area.[3]

Thinness of snow cover matters less in the southern and western regions, where winters are milder and rainfall higher, e.g. in the western Ukraine and the Crimea. Near the Crimean coast the snow normally reaches only about 6 cm. depth and lasts only ten to twenty-two days. More continental southerly regions, however, such as the north Caucasus, experience

[1] Academy of Sciences of the U.S.S.R., *Soil-geographical zoning of the U.S.S.R.* (1962) 1963, 248. This work is referred to as *S-g.z.* hereafter in this section, page numbers being to the English translation published in 1963.

[2] *S-g.z.* (1962) 1963, 224.

[3] *S-g.z.* (1962) 1963, 112.

frequent heavy frosts sufficient to ruin winter-sown crops if the snow cover is removed by strong winds.

In the dry steppes of Kazakhstan winter crops are barred by the absence of snow cover during the first half of the winter, while strong blizzards frequently remove the snow, taking soil with it. With little snow, the warming of the land in spring is delayed, and cultivation is accordingly held up, a serious matter in areas with only a short season for cultivation and plant growth.

Spring-sown crops also suffer in another way from the undue removal of snow. In the dry steppes east of the Volga, where snow cover is typically little over 25 cm. deep, extensive areas are deprived of any cover and drifts accumulate on lower land and on lee slopes, causing uneven moisture distribution and worsening the overall soil-moisture deficit.[1]

In dry areas such as the virgin lands snow contributes about a third of the annual moisture supply, so that the snow accumulation is vital to build up soil moisture against the drought that usually occurs in late spring. Shelterbelts grow slowly and measures which reduce the effect of surface winds and stabilize the snow must be taken. Cultivation which does not bare the fields of all stubble, it is claimed, can give grain yields two to four centners per hectare higher than mouldboard ploughing, and in the autumn of 1966 such cultivation was carried out on about 12·5 million hectares in Kazakhstan, Altay kray and Omsk and Orenburg oblasts.[2] However it is effected, snow retention is vital in dry-farming techniques.

Furthermore, when the snow cover is blown away or thawed by warm winds in winter and a cold spell follows, this may permit glazed frosts (*dzhuta*), which prevent grazing and cause the loss of large numbers of livestock. Throughout the steppes and wherever large areas of open farmland occur, shelterbelts are important in reducing the amount of snow blown off the soil and so increasing the amount which melts on the cultivated fields.

FROST

Some indications of length of frost-free period have been given above for certain climatic areas. More generally, the

[1] *S-g.z.* (1962) 1963, 266.
[2] A. Barayev, *Pravda*, 16.2.67, 2; *C.D.S.P.*, 19 (7), 27–28.

average period between significant frosts varies from more than 270 days on the coasts of the Black and Caspian Seas to less than 45 days in Arctic areas. The shortness of the frost-free period is a major limiting factor in agriculture in northern European areas and over almost all of Siberia, where only the valley lands average more than 90 days free (Irkutsk, 95; Yakutsk, 98). Most of the forest zone areas used for agriculture get between 135 and 150 days, rising substantially in the west (Lvov, 186). The Ukrainian steppes have 150–185 days (Kharkov, 151), but much of the Siberian steppe has less than 120 days.

The Black Sea steppes are clear of dangerous frosts by early April but the more northerly cultivated areas of the forest zone not before mid-May. Frost damage to crops occurs in about one year in three in the central areas. The Crimea is free of frosts until late September or October, and most southern areas are clear until September, but once in every three to five years damaging late frosts occur in the Ukraine. Winter frosts are sometimes sufficiently severe to damage sub-tropical plants in the favoured areas around Baku and Batumi. Over almost the whole Soviet Union frosts are either frequently or occasionally encountered during the early and late periods of cultivation so that hardly any area wholly escapes this problem.

Measures against frosts are recommended according to the nature and frequency of the threat, ranging from restriction of cultivation to covered areas in the far north and the sowing of only fast-maturing and hardy species over wide areas, to irrigation and the provision of smoke-producing equipment, particularly in southerly areas.

The extreme susceptibility of the Soviet territories to frost and drought is often overlooked by foreign critics who compare yields obtained in the U.S.S.R. with those of other lands. It must be remembered that even the hardier crops used in cooler areas are vulnerable. Potatoes, for example, are a most valuable crop in the cool, relatively moist areas of the north and west of the U.S.S.R. and are, indeed, among the few possibilities for food crops where grains do not readily ripen, but waterlogging or a few degrees of frost will kill the tubers or greatly reduce the yield.

SOIL-FREEZING & PERMAFROST

Soil temperatures are much more constant than air temperatures. Summer soil temperatures are more important than winter temperatures for agricultural purposes, but the lag of soil temperatures behind the rising air temperatures in spring delays crop growth. A substantial snow cover reduces the fall in soil temperatures and speeds up their warming in spring and so assists early cultivation and growth.

In January and February soil temperature at a depth of 20 cm. is below 0 degrees C. throughout the U.S.S.R., excepting the southern Crimea, west Transcaucasia and certain localities elsewhere.[1] In general, soil temperatures in winter and spring are progressively lower from west to east, though annual averages may be higher in the east because of much higher summer soil temperatures.

About 45 per cent. of the area of the U.S.S.R., three-quarters of the area east of the Ural mountains, suffers from permafrost. In practice, permafrost is the permanent existence of ice in the soil, cementing it together at lower levels, though thawing takes place in the upper parts (the active layer) in the summer. In the central Siberian uplands the active layer is only 20–40 cm. thick, but in the tayga of central Yakutia sandy soils thaw out to a depth of 200 cm. or more.[2]

Although soil-freezing is, in general, a handicap to agriculture, in the dry areas of eastern Siberia, permafrost conditions provide additional soil moisture to the active layer which may be exploited for hardy crops of vegetables under glass. The soil temperature in some agricultural areas of east Siberia and the Far East may fall below −20 degrees at the depth of the tillering node, but, provided snow is conserved, winter rye and perennial grasses may be grown.[3]

SOILS

The diversity of climates found in the U.S.S.R. is matched by the variety of soils (Figure 5). The close relationships between the patterns of soils and of climate and vegetation

[1] Borisov (1959) 1965, 44–45.
[2] *S-g.z.* (1962) 1963, 158, 126.
[3] *S-g.z.* (1962) 1963, 406.

5. Soil groups. (Based on *Atlas sel'skogo khozyaystva SSSR* and other sources)

were first recognized by Russian soil scientists, giving birth to the zonal concepts of soil science.

The Evolution of Russian Soil Science

V. V. Dokuchayev, during the last two decades of the nineteenth century, elaborated the need to view the soil as a complex body created and developed in functional relationship with the natural environment, as modified by man. Dokuchayev's genetic-geographical approach to soil science contrasted with that of his contemporary, P. A. Kostychev, who worked mainly on the microbiological, chemical and physical processes in soils as an aid to agriculture. Dokuchayev was also concerned with the development of agriculture, but the practical application of his ideas required the integration of the two schools of Russian soil science, which came only with the intensification of research and attention to production problems after Stalin's death. Mapping of soils was pursued throughout the Soviet Union to serve development programmes for agriculture, the needs of collective and state farms, planting of shelterbelts and other reclamation problems and the development of hydro-electric schemes.

The cartographic work led to the division of the country into soil zones, provinces, regions and sub-regions, which now provide the basis for description and further study of soils.[1] The current range of soils work is extremely wide and a comprehensive picture of the geography, chemistry and physics of soils of the Soviet Union is being assembled.[2]

In recent years, Soviet soil science has been called upon increasingly to serve the intensification of agriculture, and in particular, to guide the expansion of irrigation, the use of saline soils and the application of drainage to the problems of the areas of high humidity. This work has resulted in the evolution of meliorative soil science as a major branch of pedology in the U.S.S.R. Some examples of this work are given later (Chapter 11).

[1] *Soil-geographic zoning of the U.S.S.R.* (*in relation to the agricultural usage of lands*) was published by the Academy of Sciences of the U.S.S.R. in 1962 and translated by A. Gourevitch and published in English in 1963. As already noted, it is herein referred to as *S-g.z.*

[2] Ewald (1968) provides a convenient summary of the main fields of work and the interests of leading contemporary figures in Soviet soil science.

The zonal pattern provides the basis for the description of soils in the present chapter, but it must be stressed at the outset that the intrazonal and azonal soils are also important in an agricultural assessment of the soil pattern.[1] Indeed, the texture, depth and water relationships of a soil may be more important for cultivation or grazing than zonal features of the profile.

The texture of a soil, the depth to which weathering has occurred and the chemical and biological activity necessary for full soil development depend largely on the parent materials of the soils. Thus, under the same climatic influences, while a granitic rock weathers slowly to an acid, gravelly or sandy soil, a basalt breaks down into soil containing more clay and the clay-humus complex has a higher base status. Under the action of precipitation in a cool climate where evaporation is low, the downward movement of salts in the soil, caused by the leaching action of rainwater, results in the topsoil, in both cases, becoming increasingly acid, so that the characteristic layered profile of podzolization develops. This illustrates the development of the pattern of zonal soils in which climatic effects have subdued the differentiation caused by varying parent materials. Nevertheless, even where soils accord with the zonal pattern, agricultural activity will still reflect the parent materials of the different soil types or phases, the heavy (clay or clay loam) soils needing particular attention to drainage but responding better to selective fertilizers, while the sandy ones, despite inherently low fertility and the rapid loss of nutrients through leaching, may be valuable in conditions of difficult drainage.

The parent materials of the soils of the U.S.S.R. are very largely derived from drifts rather than from the solid rock basements. Large areas of sands, gravels and clays were deposited by ice sheets or water flowing from glaciers and these provide the mineral bases of soils in the northern and mountain areas. In the glaciated areas, the variable nature of the relief, with morainic hills and hollows, and all varieties of water conditions from overdrained rocky knolls to extensive

[1] Soils derived from parent materials which are especially influential in the development of the soil, such as limestone, are called intrazonal, while those which are still forming from relatively young deposits and have not yet weathered to a form appropriate to the climatic zone are called azonal.

marshes, results in very varied soils. Beyond the limits of glaciation, loess, derived from glacial materials, is commonly the parent material. The most fertile of the major soil groups, the chernozem, is largely derived from loess. The loessic soils are more uniform but show distinctly the effects of varying climatic conditions, and the predominant silt-sized particles are accompanied by varying proportions of sand and clay fractions.

In the north and east of the Soviet Union, the occurrence of permafrost, as already noted, is a further important factor.

Soils of the Tundra & Forest-Tundra

The word 'tundra', derived from the Karelian language, means 'non-forested places' and most of its area of about 1,700,000 sq. km. (7·6 per cent. of the area of the U.S.S.R.) is treeless, but birch, willow, poplar and other tree species penetrate far to the north in river valleys. The commonest feature of the soil types of the tundra is the presence of permafrost, which extends over virtually the whole area in Siberia and occurs in patches in the relatively small part of European Russia classed as tundra. In north-central Siberia the permafrost is hundreds of metres deep but the average thickness is about one metre, and above the continuously frozen ground is an active layer which thaws in summer. The active layer facilitates the growth of vegetation but the underlying permafrost prevents drainage, so that most of the tundra soils are marshy or peaty, with gley horizons between the peat and the permafrost. Most of the area falls within climatic region I–2, which, as already noted, is used chiefly for hunting and reindeer rearing. The better soils, when treated with fertilizers to offset their high acidity and low base status, can produce potatoes and vegetables under glass or even on sheltered open plots.[1]

In the forest-tundra near Murmansk, potato yields of 25–27 tons per hectare have been obtained, with not greatly inferior yields in other areas as far north as 68–69 degrees N. in the European areas and 70–71 degrees N. in Siberia.[2]

[1] *S-g.z.* (1962) 1963, 28, 38.
[2] Vilenskiy (1957) 1963, 254.

Grasses can be established on floodplain and tundra soils.[1] Peas, clover and other leguminous crops have given good results. The maximum possibilities are realized with combinations of peat-manuring, liming, addition of all necessary fertilizers and provision of shelterbelts, with which aids the lighter sandy loams and peat loams on southern slopes can be quite productive.

Reclamation of peat bogs has yielded encouraging results but prolonged use of reclaimed peats for grassland with regular addition of mineral fertilizers causes increased acidity, and substantial dressings of lime are essential.[2]

Podzols & Sod or Turf Podzolic Soils

The podzolic soils of various kinds cover about 11,500,000 sq. km., or over half the total area of the U.S.S.R., corresponding with the various climatic regions of moisture zones II and III and the belts of forest vegetation. From the tundra margins southward to about 60 degrees N. and eastward to the Yenisey there are various podzolic, gley-podzolic and humic illuvial podzolic soils, associated with coniferous forests developed on glacial sands and clays and interspersed with large areas of marshes. East of the Yenisey, permafrost is common and the climate is harsher, especially in the mountainous areas, so that the tayga has been less developed, but even in the European areas agriculture is found only in favourable parts, and over 60 per cent. remains forested.

The combination of cool, humid climatic conditions with climax vegetation of coniferous trees has produced the podzol. Though varying greatly with local climate, parent rock, slope, vegetation cover and other conditions, the podzols are distinguished by marked horizons in the profile. The uppermost layers (A_{00}, A_0, A_1) are dark with decaying organic matter. Leaching by slightly acid rainwater removes this and soluble salts from the upper horizons and deposits them at lower levels, leaving the eluviated (A_2) horizon strongly leached, acid and light grey in colour. The acid needles of the pine, spruce and larch trees do little to replace nutrients, and decomposition in the acid conditions is slow.

[1] *S-g.z.* (1962) 1963, 42–43, 58, 60, 63.
[2] Pereverzev and Golovko (1968).

The nutrients accumulated in the illuvial (B) horizons are largely lost to plant growth and may form hardpans which prevent free drainage even on parent materials which yield sandy soils. This hardpan must be broken up for agricultural development and large amounts of lime and fertilizers, hitherto generally lacking in the U.S.S.R., must be applied to make such soils productive.

The agricultural uses of the areas covered by these soils have been sketched in the description of climatic regions II–3 and II–4. Here we may stress that it is the soils that require no artificial drainage or very little that are most valuable. These include loams and sandy loams on valley slopes and on old floodplain terraces. Much floodplain land with poorer natural drainage may, however, be used as meadow land, and this kind of land facilitated the growth of dairy husbandry in the Northern Dvina and Onega valleys and the evolution of the Kholmogor cattle. Variations in soils are accordingly closely linked with relief and the nature of the glacial deposits that commonly provide parent materials.[1]

In Siberia, especially in the north-east, permafrost, combined with the short summer, limits agriculture to grazing, hay-making and vegetable growing under glass, with reindeer herding widespread. The moisture from the active layer of the permafrost can be used in the more favourable areas, such as central Yakutia. It is recommended that arable plots of specific size and configuration should be developed in the tayga, the shelter of the trees being also used to improve the environment for plant growth.[2]

The transition in sub-region II–3D to higher forms of agriculture was noted as occurring at about the 60th parallel west of the Ural mountains and somewhat further south in Siberia as accumulated temperatures reach 1,600 day-degrees. This is also the zone of transition to the southern tayga type of forest and the sod-podzolic soils. An admixture of deciduous trees provides a higher nitrogen and mineral supply from leaf-fall, and the herbaceous plants growing beneath the more open tree canopy, together with earthworm and bacterial activity, help to transform the upper horizons of the podzols

[1] See, for examples, *S-g.z.* (1962) 1963, 74, 79.
[2] *S-g.z.* (1962) 1963, 128.

into a turfy sod layer. The sod horizon provides better conditions for grassy vegetation and its development may be hastened by clearing the forest and putting animals to graze the land and by cropping. These soils have been studied in considerable detail.[1] The agricultural development is described in this book under agricultural regions 5 and 6, as well as under the climatic regions noted above.

Clearing the forest and establishing grazing, if not accompanied by adequate drainage and conservation, can create marshy and boggy conditions and deterioration of the podzols. Deterioration is marked by the formation of gley horizons, and if the process is not arrested peat forms and the bog moss (*Sphagnum* spp.) appears and spreads rapidly. The reversal of the process by drainage and reclamation is an important aspect of land development in the Soviet Union and is discussed later (Chapter 11).

The Grey & Brown Forest Soils

In the south of the forest zone, where broadleaved deciduous trees dominate the natural vegetation and precipitation is in balance with evaporation, grey or brown forest soils have developed. Although showing some leaching and development of horizons, these characteristics are much less strong than in the true podzol and sod-podzolic soils. In passing from areas classified as climatic regions II–4 (in the west) and II–3 (in Siberia) to III–4, forest-steppe vegetation occurs along the zone of transition, and grey forest soils are interspersed with chernozems, the grey soils having developed, in general, under the trees, while the grassland patches have yielded the black earths. Relief, however, is critical in some areas. In the Ukraine, grey forest soils and podzolized chernozems occur in the south-western, highest and most dissected areas, while typical and leached chernozems occur on flatter areas.[2] With the chernozems of this marginal zone generally leached or 'degraded', there are numerous variations in profile from the grey earth, with its more developed podzolization, through podzolized and degraded chernozems to the typical chernozems.

The profile of a grey forest soil under oak forest is, typically

[1] Vilenskiy (1957) 1963, 265–297 quotes numerous examples of this research.
[2] *S-g.z.* (1962) 1963, 209.

A_0, forest litter, dark brown and decomposing; A_1, grey, with fine roots; A_2, grey, or brown-grey, variable; B, illuvial, brown horizon, passing downward in tongues into the C horizon of brownish-grey material.[1]

The grey and brown forest soils, with their accrued fertility and moderate climatic conditions, were used from the Neolithic period and agriculture on them was relatively highly developed in the Kievan state. Exploitation for grain growing in the nineteenth century produced serious erosion, and this problem continues, somewhat alleviated by the reduction in pressure on these western areas with the addition of the virgin lands of Siberia and Kazakhstan to the grain areas of the Soviet Union. Sugar beet is important on the forest soils, but the present trend is to more balanced rotations and increases in livestock on these highly fertile but vulnerable soils.

Similar soils in western Siberia and the Altay region have been less developed and constitute usable reserves for the future. Soils of this general type also await development in the valleys of the Ural mountains, where a limited number of farms now supply milk, vegetables etc. to mining centres.[2]

The Chernozem Soils

Chernozem soils of various kinds cover 1,900,000 sq.km., 8·6 per cent. of the U.S.S.R., and provide the main agricultural resources of the country. They are associated mainly with climatic region III–4. In the north of this region, precipitation is generally adequate, given careful conservation practices, but in the south it is inadequate.

The chernozem soils were formed under natural grassland, but the presence of forests at some time in the past is argued by some authorities. Loess is the parent material over wide areas but alluvial and other sediments of weathered clays, sands, marls and limestones have also developed chernozem soils, but the manner of their evolution is still obscure.[3]

The name 'chernozem' (black earth) is derived from the A horizon, which is dark grey or almost black and generally

[1] Vilenskiy (1957) 1963, 324.
[2] *S-g.z.* (1962) 1963, 117.
[3] Vilenskiy (1957) 1963, 338–340 summarizes theories.

20–55 cm. deep. Organic and mineral nutrients are rich and the activity of soil fauna ensures mixing and decomposition of leaf and root material. There is a gradual transition to the B horizon, 40–80 cm. thick and usually of lighter brown-grey or yellow-brown colour, with calcium carbonate commonly present. The C horizon is typically pale yellow with transition to loessic parent material.

Chernozems have no markedly leached layer, though in some chernozems of the forest-steppe podzolization is distinguishable, and these soils are classified variously as podzolized, degraded or northern chernozems. Alternating with these, but on the whole occupying a more southerly transitional zone, are the 'thick chernozems' with A horizons of 40–55 cm. depth. 'Common chernozem' is the name usually applied to the soils that underlie the greater part of the steppes, the A horizon being less dark and 30–40 cm. deep. The southern steppes are occupied mainly by the 'southern chernozems', which have still lighter grey A horizons of similar thickness and a chestnut-coloured B horizon. In all chernozems there are many animal burrows and the soil may be so intermixed that horizons are almost indistinguishable.

As noted in discussing the grey forest soils, chernozems occur commonly in the forest-steppe zone associated with particular vegetation and relief conditions. In the east of the European part of this zone, between the Volga and the southern Urals, soils are chiefly humus-rich typical and leached chernozems, and heavy loams and clays predominate. Wheat is the leading crop, well suited by these conditions, with other grains, industrial crops and animal husbandry also important. In west Siberia, relief plays a greater part in this zone and marshes are widespread, but the relatively narrow, better drained, riverine stretches have leached chernozems of loamy texture, used largely for spring wheat and animal husbandry. On the interfluves, heavier meadow chernozems are used mainly for animal husbandry, with spring wheat and, less important, flax growing.[1] These examples show something of the variety, which continues eastwards through the Altay and Sayan areas, where the various types of chernozem and grey forest soils support large arable areas.

[1] *S-g.z.* (1962) 1963, 219.

In the steppe zone, chernozems dominate, though there is still considerable variety, as must be expected in this immense area, stretching from the Ukraine to the Altay and reappearing in montane basins as far as Manchuria. In the western areas, winter is relatively mild, with soils practically never frozen, but in the east winter temperatures are as low as −27 degrees C. Consequent on these and associated biological conditions the soils differ in humus content and depth of horizons, though the general profile and characteristics of the chernozem remain.[1] Wheat, maize, sunflower, sugar beet, legumes, vegetables and fruit are all suited by the physical conditions.

Poorer moisture conditions limit the use of the southern chernozems, and beyond the Volga cultivation falls to below 60 per cent. of the area, compared with over 80 per cent. in some of the ordinary chernozem areas. Sharp differences occur with variations in relief and soil-forming rocks. Drought and salinity problems increase in the Kazakh and west Siberian lands, but the cultivated areas are high percentages of total areas, particularly since the developments under the virgin land schemes.

After centuries of nomadic grazing, the more westerly chernozems were rapidly developed for grain growing late in the eighteenth century. Overcropping and lack of conservation measures accelerated erosion and depletion of the accrued fertility, and crop failures became common. The plans advanced by Dokuchayev in the 1890s resulted in some improvement, but only recently have satisfactory conservation techniques been widely adopted and the problem of erosion remains widespread in the west, and has been increased in the east by the virgin land developments.[2]

Chestnut & Brown Soils

These soils cover about 1,207,000 sq.km., 5·4 per cent. of the area of the U.S.S.R., extending from the extreme south of the Ukraine discontinuously across the south of the R.S.F.S.R., to Kazakhstan and Altay kray. The climate is classified as III–4 in the better parts, elsewhere as IV–4.

[1] *S-g.z.* (1962) 1963, 227–229.
[2] See Chapter 11.

Evaporation exceeds precipitation by two to three times, and summer rainfall evaporates completely so that autumn rainfall and thawed snow provide all the natural soil moisture.

The natural vegetation is dry steppe, with grasses and xerophytic plants, and there is relatively little humus formation, so these soils are poorer than the chernozems, though the transition is rarely abrupt. Closest in type and location to the southern chernozems are the dark chestnut soils, named after the colour of the A horizon and found in climatic region III–4. South of these are the light chestnut soils, in which the high evaporation usually produces salinification. Brown (*buryy*) soils are found in the very dry steppe regions bordering on desert conditions, and groundwater rising by capillarity brings substantial quantities of salts to the surface. Humus content falls from about 4–5 per cent. in dark chestnut to 3–4 per cent. in light chestnut and 2–3 per cent. in brown soils. Chestnut soils are weakly alkaline (pH 7·2–7·5) and low in available nitrogen, phosphates and potash, but the chief problem is dessication of the soil and of non-irrigated plants.

With the moisture deficiency overcome, these soils are quite highly productive, particularly as the climate suits hard wheats, and much of the virgin lands area reclaimed in the 'fifties was on these soils.

Serozem Soils

The serozem or grey earth soils, including the grey-brown (*seroburyy*) types, are found in the arid region of Central Asia, extending into southern Kazakhstan, with an isolated area in Azerbaydzhan. The total area of these soils and the sandy and saline deserts in the same zone is about 2,200,000 sq.km., some 10 per cent. of the U.S.S.R.

The climatic regions in which these soils occur are IV–5 (Central Asia) and III–5 (Azerbaydzhan) i.e., very warm and arid or semi-arid. Although the summers are hot, the winters are quite cold, with 80–100 days of frost, and 20–50 days of snow cover; conservation of snow meltwater is important for agriculture. Vegetation in the loessic foothill zone is varied and ephemeral, passing through a rapid cycle in spring, when there is good grazing for cattle, sheep and camels.

The most valuable serozems are in the high foothills. They

have a dark grey A horizon, relatively high in humus (2–4·5 per cent.), and a B horizon rich in carbonates, overlying the parent loess.

The greater moisture availability on the higher land gives these serozems a better structure and water-holding capacity than the lighter grey soils of the plains, where the vegetation is essentially xerophytic. Humus content of the serozems, in general, is low (0·5–2·2 per cent. on the plains and 1·5–3·0 per cent. in the foothills) and an accumulation of carbonates is characteristic. Nitrogen is low because of the poor humus content, phosphates vary with parent material, and little of the potassium content is in available form. Hence agricultural use of the serozems presents difficulties, the key to success lying in control of water.

Irrigated cultivation has been carried on in this region since about the eighth century B.C. Alongside the oases of intensive cultivation were developed livestock grazing on the plains and foothills and non-irrigated (*bogarnoye*) cropping and livestock rearing in the hills. Irrigation over thousands of years has greatly altered the serozems of the oases by the addition of sediments suspended in the water, together with manure, clay and waste material. During this century these 'anciently irrigated' or 'domesticated' soils have become largely devoted to cotton, and substantial new areas of the serozems have been developed with modern irrigation systems for this crop, as in the Golodnaya (Hungry) steppe. The ploughing of these and other virgin lands, some for irrigation, some for dry farming, has created major changes in the serozem regions, to which further reference is made in Chapter 11.

Solonchak, Solonets & Solod Soils

Soils containing a considerable amount of readily soluble mineral salts, chiefly sodium salts, cover about 750,000 sq.km., 3·4 per cent. of the area of the U.S.S.R., while salinification occurs widely in other soils, affecting at least 10 per cent. of the Soviet Union.

Solonchaks (saline, 'white alkali' or 'salt marsh' soils), characterized by a high proportion of readily soluble salts from the surface downwards, occur mainly in the serozem zone, where they cover about 22 million hectares. They have formed on various parent materials, requiring for their formation the presence of salts in the groundwater, which then rises by capillary action owing to the excess of evaporation over precipitation. This process occurs most readily in depressions, of which the largest are the Turanian and Caspian lowlands. The vegetation comprises halophytic grasses and shrubs. Agricultural improvement is possible without irrigation by introducing perennial salt-tolerating herbs, clovers and lucerne, but substantial improvement requires irrigation. Lowering of the groundwater is also necessary to stop the capillary ascent of salts to the surface, as this can result in the confluence of irrigation and groundwater and lead to secondary salinification.

Solonets ('black alkali') soils have eluviated grey upper horizons and dark compact alkaline B horizons, with salt accumulation mainly in the C horizon below 20–50 cm. They cover about 35 million hectares in the chestnut and brown soil zone in the southern Ukraine and R.S.F.S.R., Kazakhstan and Central Asia, and about three million hectares in the European and west Siberian chernozem zones. They are formed by adsorption of sodium, displacing exchangeable calcium and magnesium. This exchange of bases in a solonchak from groundwater containing sodium, or from superficial flooding, as in irrigation, produces solonetsification through successive phases of chemical reactions.[1] Amelioration of solonets soils is achieved mainly by applying gypsum during autumn ploughing of bare fallow. Irrigation assists this process, but large applications of water may introduce secondary salinification and drainage of groundwater may be necessary. With irrigation, deep ploughing can effect improvement even without the application of gypsum.

Chemical changes and the partial destruction of the mineral portion of the soil in alkaline conditions produce the solod soils found in basins in the zones of chestnut and brown soils and the chernozems. Morphologically, the solods resemble sod-podzolic soils, as under the A_1 horizon, which may be peaty, is a whitish A_2 horizon and a B horizon which is sticky when moist and very hard when dry. The upper horizons

[1] Vilenskiy (1957) 1963, 401–406.

are acidic and the lower alkaline. Improvement requires deep ploughing, liming and substantial quantities of manure and mineral fertilizers.

The Desert Lands

The fringes of the desert lands are occupied by soils of the groups reviewed above. There is a gradual transition in soil characteristics from these types to loose sand, as found in the Kara-Kum desert. Thus, solonchaks grade into takyr soils, which have a hard, almost impermeable crust, difficult to break up even with mechanical equipment and bearing only blue-green algae. Even these lands can be improved, as discovered by the inhabitants of the region over the ages, by irrigating and fertilizing.

Sandy deserts, where some water is available and little disturbance takes place, will accumulate a vegetation of extreme xerophytes, which may be grazed by cattle and Karakul sheep.

Red & Yellow Soils

The red soil (*krasnozem*) occupies only 3,000 sq.km. in two areas, western Georgia inland from the Black Sea, and the Lenkoran lowland of Azerbaydzhan, but is very important to the U.S.S.R. because of its sub-tropical produce. These are the regions with climatic classification II–5F, warm, with very high precipitation, and a natural vegetation of warm-temperate forest, under which the soils have developed on terraces cut in igneous rock.

The A horizon is of dark sepia brown with high humus content, the B horizon a vivid red or orange-red, and the C horizon dominantly bright red. A strong clay structure, high silt content and moderately acid reaction are other characteristics. The yellow soil (*zheltozem*) represents a stage towards the development of the red earth and occurs in the same and adjacent areas. Both red and yellow earths occur and are also weakly podzolized.

These soils are used for tea, citrus fruits and other crops requiring moist and warm conditions. Erosion is a constant threat and terracing is necessary on slopes of over 20 degrees or, on podzolized red soils, which erode more easily, on slopes of over 5 degrees. Organic and mineral fertilizers are required in considerable quantities.

Floodplain and Terrace Soils

Floodplain soils occupy 423,000 sq.km. or 1·9 per cent. of the area of the U.S.S.R. These soils show great variation according to the climatic region and other conditions in which they occur. Despite the hazards associated with flooding, these soils, enriched by seasonal deposition, yield useful crops of hay in addition to grazing. Soils of terraces above flood level may be used for root crops. The bottom terraces of the Volga, Dnepr, Don and other great rivers are covered with sands which require stabilization by trees and may be usable for orchards and vines.

Mountain Soils

The main groups of mountain soils in the U.S.S.R. are mountain tundra, mountain meadow, mountain podzolic and mountain steppe soils, occupying together nearly 30 per cent. of the area of the U.S.S.R., while mountain brown soils and mountain desert soils together occupy under one per cent. of the area of the Union. A variety of conditions exists on the mountain ranges of the Caucasus, Central Asia, Siberia and European Russia. Almost all these soils are under forest or meadow, but in some regions, notably Central Asia and the Caucasus, the cultivation of fruits and vegetables has been developed, in some cases over many thousands of years. Mechanized construction has made it easier to develop terraces for vineyards and orchards on the lower slopes of mountains, but there is serious disturbance to the soil profile where machinery is used for this purpose. Attention has been given to the restoration of fertility, and the study of old, manually built terraces has indicated the value of trees, such, as walnut and oak, resulting from the litter formed under them.[1]

[1] Botman (1968).

CHAPTER 5

The Labour Problem

Agriculture competes with other branches of the economy for the services of a Soviet labour force of rather more than 85 million. The population of the Russian territories rose from 159 million in 1913 to 194 million in 1940, recovered to this number, after catastrophic wartime losses of some 20 million, during the 'fifties and had increased to 239 million by 1969. During this period the rural population fell from 82 per cent. in 1913 to 44 per cent. in 1969, representing a fall in absolute numbers from 130·7 million in 1913 to 104·7 million in 1969.[1]

During the nineteenth century there was a substantial redistribution of population with continued settlement in the southern steppes and eastern territories and migration to the towns, facilitated by the abolition of serfdom. Nevertheless, such was the degree of overpopulation in the long-settled areas of the countryside that there remained a vast reservoir of labour for the new Soviet administration to draw on for industrialization during the decades that followed the Revolution. This movement continues and in some areas, at least, there is still scope for reducing the agricultural labour force and, at the same time, increasing efficiency.

Many western and southern areas still have more people living in the rural areas than are actually needed on farms. Belorussia, Moldavia, the western Ukraine, and parts of the north Caucasus, Transcaucasian and Central Asian areas are prominent among these. Some of the areas which have severe shortages of agricultural labour are not far distant from the areas of surplus, for example, the European non-chernozem zone, western Siberia and Kazakhstan, but there appears to be relatively little permanent migration from one

[1] *N.kh. SSSR 1968*, 7; *Strana Sovetov za 50 let* (1967), 15.

farming area to another, those who move preferring to seek work in the towns. The decrease in the agricultural labour force and its redistribution between the different kinds of farm are shown in Table 8.

TABLE 8

THE SOVIET AGRICULTURAL LABOUR FORCE
(averages for year, million persons)

	1940	*1960*	*1965*	*1968*
Collective farm members	29·0	22·3	18·9	18·1
Workers in state farms and other organizations*	2·3	7·1	9·1	9·4
Workers of other enterprises working on farms	0·1	0·5	0·5	0·5
	31·4	29·9	28·5	28·0
Total actually engaged in agricultural operations	28·1	26·1	25·6	24·6

* including employees of MTS and RTS, 0·5 million in 1940 and 0·4 million in 1960.
Source: N.kh. SSSR 1968, 446.

The above numbers did not include the families of agricultural workers whose contribution to farming was confined to work on personal plots, and when this labour was added the number engaged in agriculture (i.e., excluding building, repair, domestic and other workers) was estimated at 30 million in 1968.[1] The growth of reliance on state farms is reflected in the rise of their labour force from 7 per cent. in 1940 to 33 per cent. in 1969 of all those engaged in collective and state farms.

At the same time as the total labour force decreased there had to be increased dilution of the labour force with women. Even in 1969, 24 years after the end of the war, 43 per cent. of workers on state farms were women, compared with 34 per cent. in 1940.[2] In this connection, it must be remembered that males comprised only 45·8 per cent. of the Soviet population in 1968, with the shortage of men particularly marked in the generations that suffered the war as young and middle-aged adults. Also, a high proportion of disabled men added to the shortage of able-bodied male workers, while an increasing

[1] *N.kh. SSSR 1968*, 446.
[2] *N.kh. SSSR 1969*, 536.

percentage of agricultural workers approach the retiring age of 60 for men and 55 for women. Even if these folk go on working on their own plots after leaving the collective lands their efficiency must be low.

As in other societies it tends to be the more enterprising young people who are most prone to leave the farms permanently, and both single persons and young married couples find many incentives to move to the cities.

Up to 1963 the falling numbers of workers had to cope with a 45 per cent. expansion in the sown area (from 1940) without the benefit of adequate machinery or fertilizers, and this was the period of severe fluctuations in harvests both from the virgin lands, developed within the period, and from older and more fertile areas. After 1963 there was a modest contraction in the sown area, together with an improvement in deliveries of farm supplies which aided improved care of the land and livestock, and helped to offset the continuing decline in the agricultural labour force.

The labour problem was therefore somewhat less severe in the middle and late 'sixties but there are continuing problems in the regional variation of labour supply and in the availability of the more specialized and better trained workers.

As increases in production depend increasingly on efficiency and mechanization, the supply of specialists and machine operators becomes more critical. The tendency has been for specialists, such as agronomists and veterinary workers, to become available in substantially greater numbers, but for the supply of machine operators and drivers to increase at a much slower rate, as shown in Table 9.

TABLE 9

SPECIALISTS AND MACHINE OPERATORS IN SOVIET AGRICULTURE (thousands)

	1941	*1960*	*1965*	*1968*
Specialists	50	388	496	671
Machine operators	1,401	2,579	3,094	3,357
of whom, truck drivers	164	761	849	967

Source: N.kh. SSSR 1968, 448, 454.

Soviet calculations suggested that, by 1970, 2 million specialists and 10 million machine operators would be needed on the farms, but clearly this was not achieved and the weaknesses of the training system and losses of trainees to other work made it unlikely that the situation would improve rapidly.[1]

The shortage of combine-harvester and tractor drivers is particularly serious. Thus, while truck drivers increased in number by 15 per cent. between 1965 and 1968, other machine operators, including tractor and combine drivers, numbered only about 2 per cent. more in 1968 than in 1965, and a diminishing rate of increase in machine operators has been evident throughout the 'sixties.

The regional variation in the supply of specialists and machine operators is shown in Table 10.

TABLE 10

AGRICULTURAL SPECIALISTS AND MACHINERY OPERATORS
(by republics, ranked in order of numbers per thousand hectares sown land)

Specialists			*Tractor drivers and machine operators*		
	Number (thousands)	*Per 1000 hectares sown area*		*Number (thousands)*	*Per 1000 hectares sown area*
Georgia	16·7	23·0	Turkmenistan	21·0	39·0
Armenia	6·8	17·0	Uzbekistan	115·0	33·0
Azerbaydzhan	15·0	12·5	Tadzhikistan	19·5	26·0
Estonia	8·1	10·4	Moldavia	43·4	23·3
Latvia	12·4	8·3	Georgia	13·0	17·6
Uzbekistan	25·7	7·3	Azerbaydzhan	21·2	17·5
Turkmenistan	3·9	7·2	Armenia	7·0	17·5
Lithuania	15·6	6·8	Kirgizia	22·3	17·1
Moldavia	11·9	6·4	Ukraine	512·3	15·5
Kirgizia	7·7	5·9	Estonia	12·1	15·5
Tadzhikistan	4·2	5·6	Latvia	22·7	15·1
Belorussia	30·0	4·9	Belorussia	91·9	15·0
Ukraine	157·4	4·7	Lithuania	31·5	13·7
R.S.F.S.R.	310·6	2·5	R.S.F.S.R.	1,242·5	10·2
Kazakhstan	44·8	1·5	Kazakhstan	188·2	6·2
U.S.S.R.	671·0	3·2	U.S.S.R.	2,363·8	11·4

Source: N.kh. SSSR 1968, 449, 455.

[1] Wädekin (1969) 293ff.

It will be seen that the availability of specialists is highest in Transcaucasia, where, no doubt, demand for their services in connection with specialist crops is met by willingness to serve in a climatically favoured area. Next come two of the Baltic states where agriculture is relatively advanced and climatic conditions moderate, then two Central Asian republics where conditions are similar to Transcaucasia, though less attractive for natives of European areas. The other Baltic republic, Moldavia and the remaining Central Asian republics come next. Rather surprisingly, Belorussia is more favoured than the Ukraine, perhaps because of reclamation work. The R.S.F.S.R. and Kazakhstan are at the bottom of the list, no doubt reflecting the wide variety of conditions, with dominantly severe climates in the former and the low attractions of the semi-desert and virgin land areas in the latter. The R.S.F.S.R. has only one-tenth as many specialists as Georgia in relation to sown area, and Kazakhstan has only 60 per cent of the R.S.F.S.R. level.

The pattern with tractor drivers and machine operators is similar, except that Central Asia, no doubt because of the high demands of cotton, comes to the top, with Moldavia and Transcaucasia next. The Ukraine and Baltic states are in a less favoured position but above the U.S.S.R. average, with the R.S.F.S.R. and Kazakhstan again at the bottom of the table.

In an attempt to ensure for themselves the services of good specialists, many collective farms assume financial responsibility for sending promising young people away to study. In 1969, an average of 12 trainees per farm were away from one Transcarpathian district.[1] A report from Nerekhta, Kostroma oblast, stated that 124 school graduates were studying at agricultural institutes and 78 were taking correspondence courses. Greater efforts in the educational field were regarded as essential to stem the tide of rural depopulation since, despite marked improvements in living conditions and basic earnings on the collectives of 80–100 rubles per month, only 15–20 per cent. of young people leaving rural schools wanted to remain in the countryside.[2]

[1] A. Chervinskiy, *Pravda*, 15.1.69, 3; *C.D.S.P.* 21 (3), 30.
[2] G. Garnov, *Pravda*, 15.4.69, 2; *C.D.S.P.* 21 (15), 27.

In some areas the drift to the towns has become so acute as to endanger the whole agricultural sector of the regional economy. Smolensk oblast has suffered particularly and exemplifies the problem of labour shortage. The number of births in the villages in 1965 was barely one-fifth of the 1940 figure, and 96 per cent. of the pupils in the 11th class in a sample survey made among farm schools wanted to live in a city.[1] At lower levels the proportions of children who wanted to move to urban areas were lower, so, clearly, as the children drew nearer to leaving the school the attractions of the city increased, probably as they realized the necessity of leaving the locality if they were to continue their studies.

Analysis of statistical data showed that migration from the farms of Smolensk oblast began in earnest in 1950–54. At this time the farms could not offer the peasant security, and as each able-bodied member of the family, especially if single, was a tax liability they encouraged migration. In 1950, 18 young men left a settlement of 15 houses to go together to a trade school in Moscow. The limitations of personal plots, introduced subsequently, were quoted as another possible factor in driving people away from the farms.

Since then, it was claimed, conditions had so changed that, for some work at least, wages could be higher on the farms than in the towns, but young people were ignorant of this, or did not believe it. To alter the situation in Smolensk oblast, it was argued, a great social experiment was needed, with a series of integrated measures, including the advancement of qualified young people to senior posts, greatly increased housing construction, mechanization and modernization of agricultural work and the gearing of industry to serve agriculture. They stated that there had been no restriction on migration through the passport system, and completely rejected the notion that this might be so used, or that it could succeed in its objective. Clearly, they accepted that only a free worker who stays on a farm by choice will be a good worker. If choice is exercised on the factors of pay, status, amenities and other attractions, what progress has in fact

[1] G. Shinalova and A. Yanov, *Lituraturnaya gazeta*, 23.7.66, condensed text in *C.D.S.P.* 18 (33), 9–12. The survey was conducted by *Lituraturnaya gazeta* in co-operation with the Labour Resources Laboratory of Moscow State University.

been made in the U.S.S.R. to attract the farm worker to stay in the country?

EARNINGS

Throughout Stalin's industrialization drive the peasants had to suffer the burdens of low prices for their produce and high taxation. Without the exploitation of the agricultural sector the Soviet industrial advance could not have been achieved, but the return of benefits to the farmers was too long delayed. Not until after Stalin's death was any serious move made to better the lot of the agricultural workers. Malenkov and Khrushchev introduced a number of financial measures, including higher procurement prices, particularly for the less fertile areas, and the lifting of restrictions on private plots.[1] As a result, real earnings rose substantially and, in 1962, the value of a man-day on collective farms was said to be 40 per cent higher than in 1959. On state farms, from the higher base of wages, the increase was 25 per cent., but most collective farmers still received a good deal less than the workers on state farms and only 9 per cent. of collectives paid more than state farm wages.[2] Collective members' earnings will normally average less than state farm wages annually because the kolkhozniki are not employed for as many hours and there are more specialists on state farms.

The practice of paying collective farm members largely in kind, with the cash element in arrears only when the proceeds had been realized, also put the kolkhoznik in an inferior position compared with workers, including state farm employees, who were paid regular wages. Since the mid-fifties, however, there has been an increasing tendency to pay earnings in cash and in regular payments set against eventual entitlements. Guaranteed monthly payments have been supported by State Bank loans to farms since 1966.

The average value of a labour-day unit rose from 2·68 rubles in 1965 to 3·52 rubles in 1968.[3] There is, however, considerable variation in earnings both between different

[1] Outlined in Chapter 3.

[2] Lemeshev (1965), quoted by Strauss (1969) who assembles useful information on remuneration of farm workers in this period.

[3] *N.kh. SSSR 1968*, 423.

forms of work and on different farms, according to the total collective profits. The range in 1968 seems to have been 80–250 rubles per month.

On a collective farm in Stavropol kray visited by the writer in the summer of 1968 the monthly rates for a 40-hour week averaged 61 rubles in 1961, 75 in 1966 and 93 rubles in 1967. Non-cash benefits were additional. This farm appeared to be efficiently operated with good conditions for the workers but the rates quoted were low compared with some reported in the press. For example, leading farms in Brest oblast were reported in 1966 as paying grain growers 100–150 rubles per month. At the Osnezhitskiy kolkhoz in Pinsk rayon the livestock workers were reported to earn 200–250 rubles per month—and higher wages were accompanied by much higher deposits in personal accounts with the State Bank.[1]

Official figures for average wages on state farms are given in Table 11. with comparative figures for other forms of employment:

TABLE 11

AVERAGE MONTHLY MONETARY WAGES OF FARM WORKERS AND OTHER SELECTED EMPLOYEES (rubles)

	1940	*1960*	*1965*	*1968*
Average of all workers	33·1	80·6	96·5	112·6
Workers on state farms and other agricultural enterprises	22·0	53·8	74·6	92·1
Railway workers	34·2	82·9	98·7	116·0
Industrial workers	32·4	89·9	101·7	118·6
Communications workers	28·2	62·7	74·2	88·0

Source: N.kh. SSSR 1968, 555.

The monetary average wages of state farm workers remained below the national average throughout the period 1940–68 but rose from two-thirds of the national average in 1940 to about 80 per cent. in 1968. They also rose to above the level of some other low-paid groups, such as communications workers. In addition, the workers on most state farms have their small personal plots to supplement their incomes. Many

[1] *Pravda*, 29.11.66, 6; *C.D.S.P.* 18 (48), 31.

other workers also have allotments but clearly those on farms are in the best position to exploit them fully.

In addition to monetary wages, benefits in kind, such as holiday and cultural facilities, are available to most workers and these were estimated to raise the average remuneration for all Soviet workers to 151 rubles per month in 1968, i.e., by about 30 per cent. This may explain discrepancies occurring in the quotations of wages in different sources, but, in addition, there are inevitably variations between different regions and different farms within regions.

There are also different rates for different jobs corresponding to the value of the labour-day unit variation on collective farms, and average monthly wages on a state farm visited by the writer in 1968 were reported to vary from a minimum of 110 rubles to a maximum for specialists of 160 rubles.[1]

On balance, it would seem that the relative importance of private production in the individual peasant's income has fallen with the improvement in procurement prices and payment of monetary wages and pensions, all of which have provided greater incentives for the farm worker to put more effort into the collective operations. One estimate for 1958 suggested that the collective farms provided the peasants with only a quarter of their incomes, compared with three-quarters coming from the private plots.[2] This seems unlikely, particularly in view of the restrictions which had then been effective on the operating of personal plots, but collective farm members were working on average only about two-thirds of the time on kolkhoz lands that state farm workers were on the sovkhoz lands.[3] At about the same time a budget survey in Rostov oblast showed a cash expenditure by kolkhoz families about 20 per cent. above cash earnings from the collectives.[4] This may give a better idea of the relative value of private and collective earnings, at least so far as the cash element was concerned.

In 1967, it was reported, cash accounted for three-quarters of the total income of collective farm families in Stavropol

[1] See Chapter 6, p. [161]
[2] Jasny (1960).
[3] Lemeshev (1965), 266; Strauss (1969), 206.
[4] Lovell (1969), 61.

kray, 90 per cent. of this coming from the collective farm operations.[1]

Cash is not always preferred to payments in kind and there are arguments that a collective farmer may be stimulated to greater efforts if the reward is made in terms of produce. Thus, in the Kazakh republic a shepherd's entitlement to one lamb in five bred in excess of the plan was said to give good results. One shepherd raised 140 lambs from each 100 breeding ewes and received a bonus of 52 lambs. Similar incentives applied to state farm workers but in either case cash payments to the value of the produce were authorized as alternatives if the recipient preferred.[2] Combine operators who went to the virgin lands for the harvest could hand in their supplementary payments of grain at railway stations in exchange for a certificate which would enable them to collect a similar quantity—up to 14 centners—in their home districts.[3]

Some of the gap in earnings between kolkhoz and sovkhoz workers is filled by the greater produce available from the larger personal plots available on collectives. The personal plot lessens the impact on the individual of a bad year in collective harvests, while the sovkhoz worker with his steady wage needs this protection less. As the collectives now pay a higher percentage of earnings in cash, the kolkhoznik has less need to sell produce on the kolkhoz market and may prefer to use more of his small plot for basic subsistence crops. In fact, however, the proportion of private produce to be taken to market has not changed greatly.[4]

Leisure is increasingly valued as monetary wages rise and this has been cited as a reason for decreasing interest in personal plots. On a collective farm in Kalinin oblast where the size of a plot was fixed at 0·3 hectares, not everyone was making full use of it. Tractor drivers were receiving an average of 173 rubles per month and dairymaids earned 167 rubles, and did not need to go out and sell produce. Livestock, however, were still valued highly because trade in dairy products and meat was not well organized in the village.

[1] G. Zinchenko, *Izvestiya*, 6.4.69, 3; *C.D.S.P.* 21 (14), 22.
[2] *Izvestiya*, 15.6.69, 3; *C.D.S.P.* 21 (24), 24–25.
[3] *Pravda* and *Izvestiya*, 15.6.69, 1–2; *C.D.S.P.* 21 (24), 14.
[4] Lovell (1969), 62.

Fodder problems had caused a decrease in the number of cows kept privately but sheep, being easier to feed, had increased in numbers.[1]

SEASONAL PROBLEMS

Agriculture does not provide full employment for all collective farm workers, mainly because of the low farm labour requirement during the long winters characteristic of most areas. For example, in Novosibirsk oblast, in December 1967, the collective farms could offer employment to only about 67 per cent of their members, leaving 66,000, a large number of them women, without paid farm employment.[2]

Under-employment is a problem not only in the winter, when it is always acute, but throughout much of the year in most areas. So migration to the towns is desirable, but the loss of the able-bodied men causes other problems. Furthermore, at harvest time, and to a lesser extent at other seasons of high demand, labour surplus gives way to a serious deficiency in manpower (Figure 6). The only long-term solution to this situation is increased electrification and mechanization, but meantime the farms must rely at such times on a great deal of temporary assistance from workers in other occupations. The resulting recruitment of housewives, nurses, students, senior school children and other workers disorganizes their services and studies, and provides only an indifferent labour supply. Complaints about this feature regularly in the Soviet press.[3]

TEMPORARY MIGRATION

Some alleviation of agricultural employment problems occurs through seasonal migration. Soviet authorities tend to deplore this type of movement while recognizing that it is essential under present conditions. The shifting of machine operators, often with their combine harvesters and tractors, from southern areas after the harvest to northern and eastern

[1] G. Shilkina, *Izvestiya*, 21.6.69, 3; *C.D.S.P.* 21 (25), 18.
[2] Zhukovskiy (1958), 23.
[3] For examples, see *C.D.S.P.* 19 (9) 22.3.67 from *Komsomol'skaya pravda*, 1.3.67, and 19 (27) 26.6.67, from *Izvestiya*, 4.7.67.

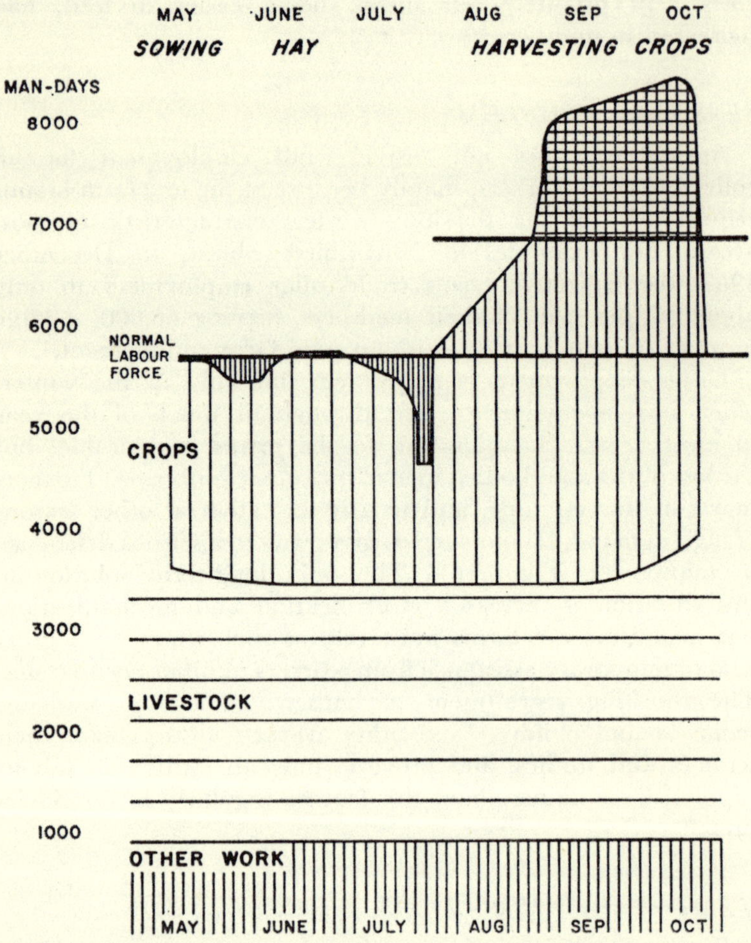

6. Demand for labour on a Siberian kolkhoz.

Seasonal variation in labour requirements poses severe problems, even apart from the lack of work available in the winter. Requirements for livestock husbandry vary only to a moderate degree but for cultivation the needs are constantly varying. The harvest absorbs all the extra hours that can be worked by the normal labour force and in the early autumn extra hands have to be recruited from colleges, schools and any other available source for the root harvest areas has generally positive advantages, and is essential for the harvests in Kazakhstan, but other seasonal migrations are of doubtful value.[1]

Apart from movements for harvesting, there are considerable movements of men in privately organized construction gangs. The movement originated in the times when poverty and the threat of starvation forced men to travel seasonally in search of work. More recently, it has provided a road to bigger incomes than could be earned on the farms, and shortage of builders enabled migrant teams to bid temporary employers up to rates far above normal, which gave the itinerant teams a bad reputation. Now, however, the rates of pay are regulated and enterprises of the Intercollective Farm Construction Trust and other bodies provide an organization within which the itinerant builders can operate. During spring and summer these men work long hours and earn high wages, meeting the needs of flourishing farms for constructional work for which their own labour resources are inadequate. They have, however, no legal recognition and no entitlement to sick pay, they do not qualify for bonuses and they live in rough circumstances while away from home.

The absence of the men from the home collectives during peak demands for labour in the fields may cause serious problems, but it would seem impracticable to keep them back for these relatively short periods when their services are in great demand for longer periods elsewhere. The magnitude of the demand is indicated by a wage bill of 80 million rubles for piecework contracts to seasonal builders in the Russian republic alone in 1965. The regional variation is considerable. Thus, in 1965, the collectives in western Siberia paid 8,500,000 rubles to outsiders via the construction organizations and Saratov oblast paid 4,600,000 rubles, but Stavropol kray apparently had no need of such work and may well have contributed valued work to the deficit areas.[2] The main need appears to be better organization of the migrant labour and, particularly, more advance arrangements so that both employing and supplying farms may know their position more securely.

[1] Wädekin (1969), 288.
[2] Yu. Chernichenko, *Pravda*, 20.11.66, 2–3; *C.D.S.P.* 18, (47), 20.

THE LACK OF AMENITIES & SERVICES IN THE RURAL AREAS

Under-employment has been noted as an acute problem on the farms at certain times of the year, especially winter. Clearly, however, individual views vary on how much there is to do in the hours free from collective work. Apart from leisure occupations, there is work in the home and with personal livestock. One kolkhoznik put it this way:

> Let's leave aside the busy periods, that's a seasonal affair. But now it's winter, and I'm back from the field where I've been hauling manure. From five o'clock, when it gets dark, until midnight, your time is your own. I am to be envied. But there's a suckling pig lying in the passage-way. It was butchered yesterday, so it has to be dressed, the meat salted, the ham cured and the gelatin prepared. And then a second hog is growing up and requires care every day. Add to that a cow, which means procuring hay and feeding and watering it. . . . Now take the house . . . we have just completed an annex and have built a new passage-way And now we're going to decorate it. . . . Parcel out all these cares to the family, and almost everyone is busy the whole evening. You get something like a second shift, the home shift.[1]

Many farms, however, are introducing two shifts in the collective work, particularly in the tractor brigades and in livestock raising. Indeed, it is reported that at the Ilyich's Behests Agricultural Artel in the Smela area in the Ukraine the milkmaids have been working in two shifts since 1952.[2]

Facilities for recreation vary greatly, depending largely on local initiative. Most collectives have a House of Culture and many have a clubhouse, but some of the peasants, particularly the younger, feel that the arrangements are too formal, or that because there are readers and students working in one room and people watching television in another they cannot talk. They would like more informal cafés. A kolkhoz, it is suggested, can easily prepare facilities for bathing and boating in summer, and for skating and skiing in the winter.

[1] Yu. Koginov, *Pravda*, 2.1.67, 2; *C.D.S.P.* 19 (1), 18.
[2] V. Fomin, *Pravda*, 8.1.67, 3; *C.D.S.P.* 19 (1), 19.

The Pravda correspondent reports:

> I arrived in Krasnoselka on a Monday. It was the day off for the local House of Culture, but people were in the collective farm's history museum and in the library and around the television set. Young people had gathered in the foyer for dancing. Many of them have their own television sets, but still they don't stay home. The habit of exchanging opinions and the desire to learn something new in talking with their comrades have their effect.[1]

A kolkhoz has to decide not only what facilities it can afford, but where they should be located. Distant brigades have sometimes been supplied with club houses before the central settlements because it is in the outlying areas that people feel most cut off.

Medical services are notoriously weak in rural areas. Although the Soviet Union has a high level of public medical care and a high ratio of doctors to population in general, the problems in farming areas which are relatively thinly populated—and steadily losing people to the towns—may be acute. Many collective and state farms have built their own cottage hospitals but this practice has been officially discouraged, it being argued that small units cannot provide adequate services.[2]

Educational facilities in rural areas are necessarily limited and the movement to the cities for higher education is inevitable and the majority who go cannot be expected to return. It has been pointed out, however, that rural areas often suffer more than they need. Some who might choose to enter a farming career are prevented by the labour legislation from starting their training at the age of 16 as, say, tractor drivers. Therefore, they continue schooling in the city and then are liable to lose interest in farming.

The failure of many collective farms to appreciate the need to offer proper encouragement to the young people to remain on the farms has also been criticized. While at 16 years, the child of a member automatically qualifies for membership of the artel, lack of formal enrolment and presentation of a

[1] V. Fomin, *Pravda*, 8.1.67, 3; *C.D.S.P.* 19 (1), 19.
[2] B. Danilov, *Izvestiya*, 15.6.69, 3; *C.D.S.P.* 21 (24), 26.

personal labour record book may result in dampening of enthusiasm and interest in farm work and eventual loss of the youngster to another job.[1]

For the young man who settles down to the rural life a further disruptive force is military service, which leads to more experience away from home and often to marriage with a city girl and the man's acceptance of city life when he is demobilized.

The low educational background of the middle echelons in farming discourages well educated young people from accepting positions under them. Of the 1,667 collective farm brigade leaders in Smolensk oblast, only 141 were reported as having secondary education, the corresponding figures on the state farms being 1,190 and 173. Most general farm workers failed to complete their schooling.[2]

THE STATUS OF THE SOVIET FARM WORKER

Collectivization has inevitably affected the farm worker's status. Before collectivization, even the wealthier peasant, the kulak, was undeniably a peasant. The poor peasant had the lowest status among workers in the old Russian empire, having advanced but little from the days of serfdom. Any change, it may be argued, which gave the worker on the land some participation in a developing enterprise such as a collective or state farm should be an improvement. Maynard examined the question of status at some length:

> The words for peasant, *muzhik*, and for peasant woman, *baba*, had nothing honourable about them. The new name for the collective farmer, *kolkhoznik*, barbarous though it may sound to the Russian scholar, has in fact more dignity.[3]

Not everyone, however, even in the Soviet Union, objects to the idea of the farm worker as a peasant. There are still traces of the sentimental attitude to the peasant that characterized the Slavophile approach in the last century, and a correspondent for *Pravda*, extolling the virtues of the peasant

[1] M. Balyayev, *Sel'skaya zhizn'*, 8.1.69, 3; *C.D.S.P.* 21 (2), 22–23.
[2] G. Shinalova and A. Yanov, *Literaturnaya gazeta*, 23.7.66.
[3] Maynard (1942), 305.

in 1966 on the occasion of Agricultural Workers' Day, 'the peasants' holiday', wrote,

> I like to use the old Russian word 'peasant', because it seems to me to have roughly the same meaning for an agricultural worker as the word 'soldier' has for a serviceman, whether an enlisted man or a marshal.
>
> Both words signify not merely professional affiliation, a standing in a given occupational structure, but a certain totality of qualities and character traits, including moral ones.[1]

Whether the farmer be considered a peasant or not, he has become a shareholder in the farm on which he works, if it be a collective, but a wage-earner if it be a state farm. If the latter, he approximates to the status of the city worker who is paid a steady wage, plus bonuses. If the former, he is dependent on the fortunes of the individual farm, but can increase his earnings, as can the wage-earner, by promotion, such as to brigade leader or to a specialist post, as well as by improving output. In either case he is assured of a retirement pension and medical benefits. Many of the older generation would probably still consider that their social status would have been higher, or that they would have derived more satisfaction from life, if they had been allowed to retain their own farms, small and marginally profitable though they might have been. As younger generations mature, however, less and less attention is given to this academic question.

The position of women is of special interest. Women provide much of the labour force on collective farms, and the work they do is largely of a manual and heavy nature. Their lot is hard, working long hours in the fields, raising children and meeting all the domestic needs of the family. Their status approximates more to that of women in less developed countries in Asia, Africa and South America than to that of women in western countries, whose occupation is normally confined to housekeeping unless they voluntarily seek outside employment for an independent source of income.

Nevertheless, some improvement on the past is visible. At least a woman worker on a collective or state farm does

[1] Y. Dorosh, *Pravda*, 9.10.66, 3; *C.D.S.P.* 18 (41), 9–10.

earn cash which is paid direct to her. Even to receive payment in kind was an improvement for a woman worker. Maynard wrote:

> The separate wage is of immense importance. I have been told, and I can readily believe, that the first actual reception of a solid dividend for the work done by the women, in solid rye and potatoes, was like the entry upon a new world. . . . A man said he had one complaint to make of the collective: *he no longer received his daughter's wages* . . . the woman's dividend is one of the reasons why there has been acquiescence in collectivization: *because it has put the women on the side of the Soviets.*[1]

Additionally, the woman gets some relief from domestic drudgery through the nurseries, schools, clubs, restaurants and field kitchens that are part of the collectivized life. She also gets paid maternity leave, short though it may be by western standards. One has the impression, travelling through the rural areas, even though less than in the towns, that the young women of today will not age as quickly as those born before the Revolution, or even those now in their thirties and forties.

The women and young people generally have also gained freedom from the tyranny of the head of the household. Drunkenness and wife-beating may still occur, but the woman has some redress at hand from the officials of the collective if social disapproval is insufficient.

In the Soviet system, age and sex alone do not determine authority. Though the men still dominate government on both the national and the local scale, women have more influence than formerly. Women are rarely given the highest posts but many become specialists or brigade leaders, besides achieving distinction and high earnings tending dairy cows, pigs and poultry.

Shimkin discussed six causes of inadequacy and tension contributing to the problems engendered by the politico-economic demands of the government.[2] Besides the misuse of labour, he noted inadequate financial returns to farm labour (somewhat eased since he wrote); inadequate labour

[1] Maynard (1942), 314.
[2] Shimkin (1963), 86–92.

productivity, associated particularly with health problems, which are greater in rural areas than in the cities; rural class stratification, including contempt for the rural labourer by city people appointed to farm management and other senior posts; and ideological tensions between Communist Party members and conservative elements in the countryside. These include differences over traditional practices, such as survival of polygamy, or attempts to revive it, child brides, bride prices and marriage by abduction in Central Asia, religious observance and messianic movements, opposition to Communist practices and friction over private plots. Finally, individual administrative mismanagement, despite generally high levels of personal integrity, causes frequent complaint in the Soviet press.

Shimkin found evidence of transition to a modern farm society in some areas, notably the eastern grainlands of Siberia and Kazakhstan, but, in contrast, he noted the survival of peasant characteristics in the south-west, Vyatka and Transcaucasian regions. He also reported serious tensions in the rural areas of the northern industrialized zone and latent in Central Asia and the Transcaucasus.[1]

In conclusion, it seems clear that the lot of the farm worker has improved considerably both in regard to remuneration for work performed on the collective lands and in prices obtainable for the produce of his personal plot. There is, however, no evidence to suggest that the worker is better off financially on a farm than he would be if he changed to an urban job and there are many pressures of a non-monetary kind to persuade him in this direction.

[1] Shimkin (1963), 92-101.

CHAPTER 6

The Individual Kolkhoz and Sovkhoz

The foregoing chapters have said little about individual farming units. This chapter will show that considerable variation exists not only in the land-use patterns on different farms, differences essential to meet climatic, soil and other conditions, but also in organization.

Broad policy lines are laid down centrally and the regional committees no doubt exert pressure towards conformity, but there still remains sufficient autonomy for the individual collective farm to plan for somewhat different economic and social development from its neighbours, and to make limited experiments in its use of labour and other resources.

Indications of this moderate degree of freedom, already noted, are the variations in payment of earnings—some collectives paying wholly monetary wages, others part in cash and part in kind—as well as in the rates paid for similar operations. State farms vary less in this respect, being essentially on a wage basis, with wages less closely related to farm earnings, but there is still scope for variety, particularly since the transfer of state farms to a self-supporting (*khozraschet*) basis.

An example of experiment by a state farm was the reorganization of cultivation methods on the link system in the Labour Sovkhoz in Volgograd province in 1966.[1] The account of how the experiment was implemented illustrates the chain of control. The report began with the director of the Mikhaylovka rayon agricultural administration requesting an appointment with the chairman of the oblast executive committee. He arrived with the director of the farm and a section manager. The farm was concerned about productivity of men and machinery in the grainfields, the average area tilled per tractor unit in the oblast being only 150 hectares.

[1] V. Pokrovov, *Pravda*, 30.8.66. 2; *C.D.S.P.* 18 (35), 28–29. Reference was made to the link or *zveno* system in Chapter 1.

The section manager had come to the conclusion that productivity could be improved by giving tractor teams responsibility for particular parcels of land—instead of one man or team doing the autumn ploughing, another the preparation for sowing, another the actual sowing, with yet others gathering the harvest, one team should have responsibility for one area throughout.

The agreement of the rayon and oblast committees having been obtained, the ploughed land in the experimental section was divided into ten areas of 411 hectares each allocated to the teams by lots. Eight machine operators decided to work in pairs, the remaining two decided to work separately. The sovkhoz continued to provide all inputs and would, of course, market the produce, but incentives were provided by offering increased earnings. Thus, one of the men working on his own was told that his average monthly payment on account would come to 201 rubles with a bonus if his plan was achieved, and for grain over and above the plan he would receive an additional 20 per cent. of its surrender value.

Initial results were most encouraging. The machine operators on the experimental section harvested in the first year an average of 378 tons of grain, valued at 22,700 rubles, compared with 118 tons worth under 7,000 rubles for the average of the whole farm. It was decided to put the whole farm under the new form of organization—not surprisingly, if these figures are for production under otherwise comparable conditions. Other consequent economies emerged. For example, whereas each section agronomist was acting largely as an administrator, allocating machinery and jobs, it was suggested that the farm agronomists should pool their talents and each concentrate on the aspect of husbandry to which he was best suited, such as studying tillage methods, sowing practices, rotations etc.

While press reports give some insight into the workings of individual Soviet farms, they never provide comprehensive reports. Technical coverage is made available through the media of special reports for agricultural scientists but these are invariably concerned with farms involved in experiments, and, even in these cases, it is usually the experimental section rather than the farm as a whole that is described.

Similarly, it tends to be the rather exceptional farm to which foreign visitors are taken. It is virtually impossible for the foreigner to obtain permission to visit farms of his own choice and there is absolutely no question of visiting a random selection of farms. It is, however, possible to arrange visits locally to some extent and thereby to see farms which are not specially organized for visits from tourists. This appears to be as near as the outsider can get to objective assessment, but it does enable something of the character of a farm, its organization and its economy, to be reported abroad. It is suggested that such farms will not include poor examples, but may be representative of good farms, not necessarily only those distinctive in their area.

Detailed accounts of a number of Soviet farms were given by Dumont following his tour from Moldavia to Kazakhstan in 1962.[1] These accounts deserve comparison with his earlier descriptions, which were not based on his own visits, but which depicted sample farms at an earlier stage of development.[2] The later accounts, also, however, have necessarily become somewhat dated, but their detail makes them still of special value.

In 1968 the present writer visited a number of collective and state farms and the following accounts are based on these visits.

THE UKRAINKA EXPERIMENTAL SOVKHOZ & THE KUTUZOVKA SOVKHOZ, KHARKOV

Attached to the Animal Husbandry Research Institute of the Forest-steppe and Wooded Lands of the Ukrainian Republic near Kharkov, are two large state farms specializing in livestock husbandry, particularly in improving the breeds selected for planned development in the Ukraine. High grade pedigree livestock are bred and sold to collective and state farms and to pedigree livestock centres.

Both pure-breeding and cross-breeding of cattle, sheep and pigs are undertaken, and also some breeding of horses and rabbits. The full support of the Institute's scientists—117 research workers and 60 post-graduates—is available to help

[1] Dumont (1964).
[2] Dumont (1954) 1957.

solve technical problems, but the farms are run commercially and make substantial profits from the sale of pedigree stock, milk and other produce.

Virtually the whole cultivated area of the farms is devoted to the livestock enterprises, most of it to fodder production. The areas under different crops provide an interesting summary of what the Soviet agriculturalists call the 'green conveyor system', by which a succession of different fodder crops ensures continuity of supplies of feedstuffs. Naturally, the actual crops used vary from one area to another according to climatic and soil characteristics.

In this area of the wooded steppe zone, precipitation totals about 400 mm. per year. Water shortage is a perennial problem, though rarely acute. The fields of 100–150 hectares on these farms are divided by forest belts which provide wind protection and facilitate snow retention, so increasing soil moisture. There is some soil erosion by wind and water but only locally is it serious, mainly on slopes where rills and gullies are liable to develop. The soils, of chernozem types, with careful use and moderate applications of fertilizers, are very fertile and, with water conservation also given attention, a wide range of crops and grasses can be grown successfully. Yields (centners per hectare), are, winter wheat, 35–40, maize, 45–50, lucerne hay, 50–55, maize for silage, 350–400 (green mass), fodder beet, 450–500 (green mass).

Substantial portions of the grass and grain crops are dried and powdered, so less hay is baled than formerly, but 333 hectares of the 'annual grasses' and 233 hectares of the 'perennial grasses' were for hay in 1968. There is, of course, continuous adjustment of cropping. Thus, since 1961, although the sown area of Kutuzovka farm has been increased from 2,985 hectares to 3,138 hectares, various changes have increased the fodder resources. Potatoes and vegetables have been cut from 272 to 10 hectares, being now grown only for internal domestic use. The areas under wheat, maize and peas have been moderately increased from only 27 hectares in 1961 to 200 in 1968. They had, however, in the previous few years been somewhat reduced from their peak in favour of green fodder crops, which now occupy 53 per cent. of the area, compared with 45 per cent. of the smaller area in 1961.

The cropping on the two farms in 1968 was:

	Ukrainka hectares		Kutuzovka hectares	
Grain:				
Wheat (winter)	506		412	
Maize	870		400	
Barley	594		200	
Peas	521		200	
Oats and others	331		137	
TOTAL GRAIN CROPS		2,822		1,349
Sunflower	148	148	110	110
Potatoes and vegetables	26	26	10	10
Fodder crops:				
Maize for silage	1,244		824	
Fodder beet	201		110	
Perennial grass, (including lucerne)	552		212	
Annual grass (including Sudan, and vetch-oats mixtures)	1,115		498	
Misc. (green winter crops, melons, etc.)	117		35	
TOTAL FODDER CROPS		3,229		1,679
TOTAL SOWN AREA	6,225	6,225	3,148	3,148
Pasture, etc.		778		508
TOTAL AGRICULTURAL AREA		7,003		3,656
TOTAL AREA		7,634		4,082

This pattern of land use is unusual for this region because of the dominance of the livestock enterprises. Most farms here have more land under winter wheat and 10–25 per cent. under sugar beet, with green fodder crops occupying 28–33 per cent. of the sown area. Many have up to 10 per cent. of their crop land fallow, which the Institute regards as unnecessary in this zone, given satisfactory rotations and high rates of fertilizer application. Further south, irrigation would be needed if fallow were to be dispensed with.

Livestock numbers, at 1st January 1967 were:

	Ukrainka	Kutuzovka
Cattle	3,528	2,896
including cows	1,287	1,100
Pigs	4,422	1,280
including sows	400	100
Sheep	4,281	—
including ewes	1,000	—

The Ukrainka farm has herds of several breeds—Simmenthal, Lebedinskiy, Red Steppe, and Black Speckled (Friesian type). Cross-breeding has introduced many strains, including Shorthorn, Hereford, Aberdeen-Angus and Charollais. The beef qualities of Red Steppe and Simmenthal have been much improved in this way, and a new type of Red Steppe, yielding milk with a 4 per cent. butterfat content, has been evolved. The Grey Steppe breed has also been improved here, and the Lebedinskiy breed was largely evolved in the Institute.

The Kutuzovka herd produces whole milk for sale in Kharkov and is based on the Friesian type. Average milk yield in 1967 was 2,905 kg. and it was hoped that this would be raised to 3,250 kg. in 1968. The highest yields, about 4,500 kg., were obtained from cows in the fourth and fifth lactations, falling by about 10 per cent. in the 7th lactation. Butterfat content averaged 3·75–3·82 per cent. The labour input was 2·4 man-hours per centner of milk, and the total cost 10 rubles 90 kopeks, giving a profit of 35–40 per cent., considerably higher than the average farm would net because of the high quality herd and superior organization. Labour is strictly specialized—for example, milkmaids are concerned only with milking—and there is a high level of mechanization, including delivery of feed by tractor. The cows are kept in units of 400, divided into groups according to productivity, live-weight, condition, etc., and each cow's record is kept in considerable detail. They are kept untethered, and their exercise involves walking about 4 kms. per day, but they are not pastured.

Calves when weaned are put into boxes of 15 for controlled feeding with not more than ten days' difference in ages. Milk is delivered automatically with a 3–4 minute feeding time for each group. Young cattle are vaccinated against foot-and-mouth disease at 6 months and one year old, with repeat vaccinations later, but not every year.

The milking is carried out by three 'Yelochka' units which permit 56 cows to be milked simultaneously. One milkmaid in 1967, expertly managing the preparation, feeding of concentrates and actual milking, obtained 3,800 centners of first grade milk, the average being about 3,000.

The milk is cooled and pasteurized before dispatch.

Pig Breeding

The principal breed is the Large White, but Mirgorod (a black mottled breed named after the town half-way between Kharkov and Kiev), Landrace and Welsh breeds are also used. The latter two are particularly valued for the improvement of the meat qualities they impart.

The pigs are housed by age groups. Measured quantities of feedstuffs are delivered to the bays by tractor and manure movement is also mechanized. The bays are designed to prevent a sow lying on her piglets, which average nine or ten. The houses are air-conditioned and ultra-violet lamps are used. The pigs are allowed out twice per day in suitable weather and are provided in the yards with extra feeds of beetroot. Piglets are allowed out in summer after two weeks. Piglets weigh 20–21 kg. at eight weeks when they are weaned.

Pork is the main object, with slaughter at a heavier weight than is usual in Britain, about 95–105 kg., reached in 150–180 days by the better breeds and up to 250 days by others. Bacon pigs are killed at about 85–90 kg., but lard pigs (for which demand has declined) are grown to 150 kg.

Sheep

Sheep breeding is concentrated at the Ukrainka farm and concerns the development of the Prekos and Sokolskaya breeds, particularly the former, which has considerably improved the local Ukrainian sheep. It is a semi-fine-woolled breed. The average fleece weight is 4·5 kg. but a good Prekos ewe will yield 7 kg. of 8 cm. staple and a good ram over 10 kg. of 9·5 cm. staple. The Sokolskaya is a much lighter breed, a good ewe having a live-weight of 50 kg. compared with over 60 kg. for a Prekos, and the fleece is very light. Improvement of the Karakul breed, again small but valuable, is also in hand. In conjunction with the Stepok stud farm, the Ukrainka farm has produced a new dual-purpose cross of Prekos and Askanian breeds. Lambing rates vary with breeds but average 120–130 per cent.

The numbers of staff and workers of all grades on the farms were, in 1967, 1,200 on Ukrainka and 450 on Kutuzovka. Work on Ukrainka is organized in 13 brigades (4 cultivation,

9 livestock) and on Kutuzovka in five (2 cultivation, 3 livestock).

Wages depend on production but average monthly earnings were, in 1967–68:

	rubles
Livestock workers	110–120
Skilled mechanics	140–160
Specialists	150–160
Director	200–220

Accommodation is largely in flats so there are no garden plots as such, but allotments of up to 0·15 hectare are available near the flats and on these some workers keep pigs, but not cows.

All transport is mechanized and there are 75 tractors. Aircraft from a regional centre are used for spreading fertilizers, herbicides and pesticides when aviation is judged to provide the best means.

Fodder conservation and movement is also highly mechanized. Experiments have been made with both tower and pit silos and the latter have been found to be more economical. The pits are constructed to hold 2,500 tons each. The silage is covered with plastic. The base may be used as a threshing floor. Harvester towers, holding mixed hay and silage are also being used.

The artificial insemination station attached to the farms is claimed to be the world's largest. It deals with cattle, pigs and sheep. About 150 pedigree bulls of 8 breeds are kept and their semen is frozen down to −196 degrees C. for long-term storage. Collective and state farms are sent 1,000 doses, about a year's supply, in one large thermos-type container. Liquid nitrogen is sent at intervals of 20 days in summer and 35–40 days in winter for temperature control. Ram's semen will not withstand deep freezing but will keep for 48 hours at freezing point so dispatches are made to farms on the days required.

THE KOMMUNISTICHESKIY MAYAK (*Communist Beacon*) *KOLKHOZ, STAVROPOL KRAY*

This large, apparently wealthy, collective farm is situated about 80 km. east of Pyatigorsk, with an outlying settlement

110 km. further east, adjoining the Dagestan A.S.S.R. It is essentially a steppe farm, the main area, around the settlement of Kommayak, being near the southern edge of the steppes. The land here rises gently to the foothills of the north Caucasus, but is nearly flat. With an annual rainfall of 250–300 mm. it is almost all cultivable, though the chestnut soils suffer from salinity. The outlying area, Buruny, is more arid. About one-quarter of Buruny is semi-desert; the rest has sandy soils, suitable only for pastoral use and, in the better parts, for hay.

Kommayak is linked to Georgievsk and Pyatigorsk by fairly good straight roads, nearly all tar-sealed. Georgievsk, 45 kms. from Kommayak, is a railway junction where a branch line connects with the main lines from the Caspian ports and Ordzhonikidze to Rostov-on-Don and Moscow, so the farm is well placed for transport.

The nucleus of the collective farm was a commune established in 1920. This included 48 people, who together owned 82 hectares of land, 20 working cattle, 16 horses, 12 cows, 32 sheep, 8 ploughs and 3 mowing machines. The high proportion of the livestock kept for draught purposes indicated the large amount of fodder needed for the maintenance of animals not directly for food. This dependence on animal power began to lessen in 1921, when the commune acquired its first tractor on credit from the state.

The creation of the collective farm from this commune and surrounding areas took place in 1934. On the eve of the Second World War it occupied 12,000 hectares, and included 1,017 people, with 280 cows, 11,000 sheep and 400 pigs. Sheep predominated because much of the land was not cultivated. The area was occupied by German troops from September 1942 to the 1st January 1943 and suffered severe losses in lives and material goods. Reconstruction and development have cost about two million rubles. The last major change took place in 1962, when another collective was added to complete the present unit.

The collective now has a population of 2,500 people, of whom about 670 men and 490 women form the labour force. The total area of land is 21,700 hectares, of which 12,800 are in the main block. The Buruny block, 110 kms. distant, covers about 9,000 hectares, of which about 1,000 hectares are cut for hay. The total agricultural area is 19,728 hectares, of which 10,300 hectares are ploughed.

The structure of the ploughed area was, in 1968:

	hectares		*per cent.*	
Winter wheat	3,833		37·2	
Barley	400		3·9	
Maize (for grain)	400		3·9	
Oats	200		1·9	
Millet	400		3·9	
TOTAL GRAIN CROPS		5,233		50·8
Sunflower	1,003		9·8	
Coriander	200		1·9	
TOTAL OIL CROPS		1,203		11·7
Perennial grass, incl. lucerne and esparto	567		5·5	
Annual grass, incl. Sudan	673		6·5	
TOTAL SOWN GRASS (for hay and grazing)		1,240		12·0
Maize, for silage	1,100		10·7	
Winter green crops (wheat/barley)	600		5·8	
TOTAL GREEN FODDER		1,700		16·5
Melons	87		0·8	
Other vegetables	14		0·1	
TOTAL VEGETABLES		101		1·0
Fallow (wheat land)	823	823	8·0	8·0

In addition there are 400 hectares of gardens and orchards, comprising 200 hectares of fruits, including apples, cherries and apricots, and 200 hectares of vineyards.

Livestock in 1968 were as follows:

Cattle	
Dairy herd: (including cows and heifers in calf but not in milk)	847
Bulls	6
Other cattle, over 18 months old, male	313
Other cattle, over 18 months old, female	288
Other cattle under 18 months, male	535
Other cattle under 18 months, female	498
TOTAL CATTLE	2,487
Sheep	
Ewes	5,273
Rams	43
Wethers	4,577
TOTAL SHEEP	9,893

In addition to these main livestock enterprises the farm keeps about 90,000 poultry, including 30,000 hens for egg production, and a small number of pigs. The collective is also interested in horse breeding and has about 360 horses.

Management & Operation

The management of the collective consists of the chairman and a board of nine members elected by a general assembly of the collective and serving for three years. The staff of specialists includes 12 agronomists, 10 animal husbandry specialists ('zootechnicians'), 7 veterinary officers and 10 machinery specialists.

The work on the collective fields and livestock is delegated to 15 brigades, having the following responsibilities:

	No. of brigades	
Field crops	4	
Cattle	3	
Sheep	1	
Poultry	1	
Pigs	1	
Horse breeding	1	
Orchards	1	plus one specialist tractor brigade
Vineyards	2	

The main items of machinery are 70 tractors, 19 grain combines, 38 vehicles and 239 electric motors of various kinds. The farm has an 'Agregat' for drying grass and grain. Machinery repairs which cannot be effected in the farm's own workshops are performed by the agricultural technical service (*Sel'khoztekhnika*), which also supplies fertilizers, mainly phosphates, nitrogen and potash (especially for the sunflower crop). The state grids supply electricity and gas.

In the basic rotation, land is fallow for a year, after which two crops of wheat are taken, and then lucerne is sown, but there are, of course, many variations to this pattern. The variable incidence not only of precipitation but also of the hot dry winds from the east, causes frequent failure or partial failure of crops, and yields of wheat vary from 10 to 26 centners per hectare.

The three milk-producing units all have Red Steppe cows. One herd, comprising 230 pure-bred cows, is essentially for dairying. The 452 cows of the other two herds produce Charolais-cross calves to be sold for meat at about 18 months old. Milk yields average 2,500–2,700 kg. per lactation.

The succession of main constituents in the livestock fodder illustrates the diversity of crops needed to ensure continuity of supplies:

Spring	(May)	Winter crops—barley
Early summer	(June)	Oats
Mid-summer		Sudan grass
Late summer	(August)	Maize
Autumn	(Sept./Oct.)	Sudan grass (second growth)
Winter	(November)	Silage

The sheep-breeding enterprise, located at Buruny, is based on the Merino breed for fine-wool production and cross-breeding for meat and wool. The sheep are kept in regular age groups, the 1968 figures being: one-year ewes, 963, two-year ewes, 1,061, older ewes, 3,249, wethers under two years, 2,208, over two years, 2,369. Three flocks are pure-bred with wool as the main object, while in two flocks the Merino ewes are crossed with North Caucasus rams to improve their meat qualities. Average wool yield is 5–6 kg. Artificial insemination is used for sheep as well as for cattle.

The poultry enterprise, as already noted, maintains about 90,000 birds, of which one-third are laying hens. Average production is 170 eggs per bird. Broilers are also produced.

Some of the output is sold to the co-operative organization, SELPO (*Sel'skoye potrebitel'skoye obshchestvo*), but the state purchasing organizations are the main outlets for the collective produce. This comprises grain, milk, meat, eggs, wool and some of the grapes and other fruits. Wine is, however, made on the collective, annual production being about 15,000 decalitres. Direct sales of some of this wine and fruit, vegetables and flour are made through shops in which the farm has an interest. The collective has a cold store, a flour mill, a bakery and an oil factory as well as the winery.

Of the income of the farm (about 2·5 million rubles per year) 53 per cent. is derived from crops, 41 per cent. from livestock and 6 per cent. from miscellaneous sales. Earnings have been paid wholly in cash since 1958. A work-day of 8 hours earned an average payment of 4 rubles 53 kopeks in 1967, compared with 3·68 in 1966 and 2·98 in 1961. A year's work of 245 man-days would thus have earned 1,110 rubles in 1967 compared with 905 in 1966 and 730 in 1961. The supplementary value of the culture-fund benefit is estimated at 170–180 rubles per year. Nurseries are free and there are a music school and a ballet school for children as well as a library, a wide-screen stereophonic cinema and clubrooms in the *dom kul'tury*, and sports fields and a lake for swimming and boating. There is a 35-bed hospital on the collective.

All the 600 houses of the farm have piped water, electricity and gas. A typical house, visited by the writer, was single-storey, built on a square plan, with porch, kitchen, store-room, living room and two bedrooms each with two beds, one having also in it a desk, bookshelf and sewing machine. All rooms were neatly decorated with plain but good furniture. The living room contained a television set and a radio. The house was centrally heated from a gas plant housed in a garden shed which also contained tools.

Each house has a garden and personal plot of 0·15–0·30 hectares for which a rental of 12 rubles per year is paid. The houses are the private property of the occupants, and can be sold if the owner leaves the farm or wishes to sell for any other reason. A cow and a calf, pigs, bees and poultry may be kept on the personal plots, which also produce vegetables, grapes and other fruit and flowers. The household visited by the writer did not keep livestock, the owners saying that this was unnecessary as they obtained their milk and meat from the collective. All members of the collective can sell produce through the co-operative organization.

In several years the farm has failed to fulfil the plan received from the rayon agricultural committee because of unfavourable weather conditions, but 1968 was progressing favourably when the farm was visited in August.

Current farm development plans include connection with a reservoir 16 kms. distant to provide water for irrigation, especially of grain and grass crops. This, it is thought, would largely eliminate poor yields in dry seasons.

THE PO ZAVETAM IL'ICHA (*Legacy of Ilyich*) *KOLKHOZ, NORTH OSETIAN A.S.S.R.*

The settlement of Arkhonskaya, near Ordzhonikidze, is the functional centre of this collective farm, which is probably typical of an average collective. Its sown area of 3,400 hectares compares with the U.S.S.R. average (1967) of 2,800 hectares, and it is clearly less wealthy and less advanced than some to which overseas visitors are frequently taken. Development, is however, taking place rapidly and the appearance of the farm will have changed greatly in a few years.

Arkhonskaya is on nearly level land at the foot of the Caucasus mountains, and the collective owns forest and pasture land in the foothills, 20 kms. distant. It is, however, primarily a farm of the Terek valley and has ready access to the railway and trunk-road systems at Ordzhonikidze.

The total area of the collective, including the mountain pasture and forest lands, is 7,000 hectares. The cultivated area is 3,600 hectares, of which 200 hectares are orchards. Grain crops predominate in the sown area, with about 700 hectares of winter wheat, 400 hectares of barley, 400 hectares of maize for grain and 300 hectares of buckwheat. Some 400 hectares are devoted to vetches and oats, and 250 hectares to lucerne and 100 hectares to clover—both classed as perennial grass. Other fodder crops include maize, harvested when not fully ripe, rape and grass. Fodder is mainly dried grass and grain, and silage. 12,000–15,000 tons of silage are made annually. The drying of crops and their powdering for fodder is a new feature on this farm, which received a machine for this only in time for the 1968 crop.

In addition to the grain, crops grown for sale include potatoes, sugar beet, hemp, and vegetables, especially tomatoes and cabbage. Fruit, especially apples, constitutes an additional source of income. The farm is, however, investing heavily in its livestock branches, which are dairying and the rearing of about 7,000 pigs. Two large units have been constructed for intensive stall feeding of dairy cows, each holding 140 cows, with automatic drinking bowls and mechanized feeding and cleaning arrangements as well as electric milking. The cows, of the Red Steppe breed, are in the stalls throughout the year, except for yard exercise for two hours per day. A staff of four is attached to each unit so eight persons care for the 280 cows in these units. A milk processing plant was being built on the farm in 1968 and was expected to increase receipts from milk sales by 50 per cent.

The farm had 58 tractors, 19 grain combines and 55 motor vehicles in 1968, a rather high total for the size of the farm, and much of this equipment may well be obsolescent, a suspicion supported by the retention of about 100 working horses. Electricity is drawn from the grid, to which the farm was connected in 1964.

The total number of families associated with the collective is about 1,400, a high number for a collective of this size. About 150 hectares are occupied by the garden plots attached to the houses, with a similar area allocated to separate allotments. Some areas of poor, uncultivable land are allocated for grazing of personal livestock.

New building in the kolkhoz includes two-storey, semi-detached brick houses. Most of the older houses are single-storey. A new administrative centre was also being built to house the chairman and staff and provide committee rooms and a large hall.

CHAPTER 7

Grain Farming

Grain production has been the focal point of Russian and Soviet farming throughout modern times. Grain and other products of the arable land have been far more important in the food of the people than have livestock products. Only in the peripheral regions of mountainous, semi-desert and other inhospitable lands have livestock been accorded more attention. The situation is changing somewhat with higher standards of living and consequent increasing demand for more varied food products, but the switch in land use is more from direct food crops to fodder crops than from arable to grassland farming.

Hence, the shortage of arable land remains critical in Soviet planning of land use. Of the total land area of the U.S.S.R. of 2,227·5 million hectares[1] only 608·1 million hectares are classed as agricultural land, and little more than one-third of this is arable (Table 12). Furthermore, about nine-tenths of the

TABLE 12

AGRICULTURAL AREA OF THE U.S.S.R., 1969
(million hectares)

	Arable	*Hay*	*Pasture (excluding reindeer pasture)*	*Total agricultural land*
Land used by farms*	223·3	40·6	274·1	546·2
State land reserves and forest organizations and other land holders	1·0	6·9	53·3	61·9
Total agricultural area	224·3	47·5	327·4	608·1

* Including subsidiary agricultural enterprises.
Source: N.kh. SSSR 1969, 304.

[1] The Azov and White Sea areas are excluded.

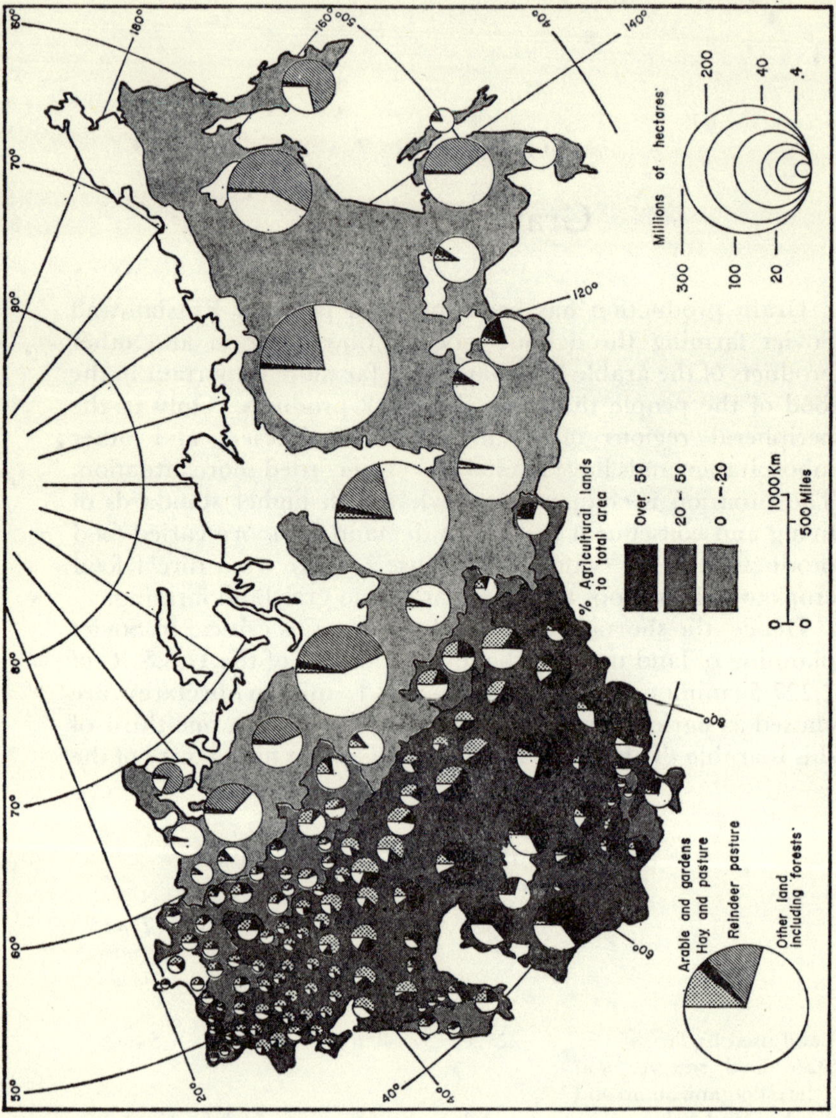

7. Major classes of rural land use, 1965.

arable land is found in the south-eastern quarter of the territory and even here large areas are unsuited to cultivation. This uneven distribution of land resources in terms of use-capability is illustrated in Figure 7. The divided circles show the proportion of the total land area of administrative units grouped in each main class of land use and the percentage of land devoted to forest, reserve or other non-agricultural uses is shown in white, a significant proportion in all areas. Pasture uses are relegated to the poorer lands but also occupy large areas. In Central Asia, particularly, pasture quality is low because of the aridity of the climate. In the north-east, most of the land is classed as reindeer pasture, forest or scrub.

In the whole of the U.S.S.R. then, about 224 million hectares are classed as arable (*pashnya*), roughly one hectare per head of the population. The area available for crops each year is reduced by the necessity for fallowing a proportion of the land in the drier areas.

Deducting the area of fallow from the total arable area gives the area sown to crops, including sown grasses, (*posevnaya ploshchad'*) and here referred to as 'total sown area'. Strictly, however, the areas reported as sown appear, at least sometimes, to be the areas from which crops were harvested, including grazed temporary grass, rather than original sowings. Hence, if an area is sown with winter wheat and this is destroyed by frost and the land resown, it is the latter sowing which appears in the final statistics. There is only occasional mention in the press of areas originally sown, or of differences from final areas of crops, so the term 'sown area' as used here means the area reported as such.[1]

Rotation of crops is now normal in Soviet farming, though, of course, there are crops which are suitable for monoculture. In their rotational practices, Soviet agriculturalists speak of a 'field system', such as an eight-field rotation or a six-field rotation, rather than the number of courses through which a given field is passed as in English usage. Though this is at first confusing to the observer from the west it has some merits. A brigade's fields in a given year are allocated to their respective crops, but some crops are in the ground longer than others, so that there are sub-stages in the rotation in some fields and not in others.

There is much discussion on rotations and sometimes there is

[1] The position is, of course, similar in most countries where the annual census records the farmer's figures of his crops actually in the fields at the time of census. The difference is that the term 'sown area' is not generally used.

pressure from Party or other circles to conform or approximate to particular rotational systems, the best known instance being Stalin's espousal of the Vilyams *travopol'ye* system, discussed on pages 56–58 and 208–209. There is usually some politically desirable crop—such as maize when Khrushchev was urging it as a solution of fodder troubles—and resulting pressures interfere with the considered judgments of farm managers and agronomists. Administrators and advisers at rayon and oblast level may also influence farm practices and, reinforced by pressure from above, may virtually dictate rotations, at least to the weaker farms. This is not admitted officially, the 'party line' being as stated in the following quotation: 'In present day conditions, the problem of increasing the output of grain, industrial crops and fodder is most successfully solved by the field-crop rotation of fallow-intertilled and fallow-grain crops. Guided by the state order-plan for agricultural output, the collective and state farms themselves determine, without any imposition whatsoever from above, the most advantageous structure of sown areas.'[1]

No outsider can unravel these pressures and relationships. It is the writer's impression, derived from visits to collective and state farms and press comments,[2] that the farm management is freer to decide its own practices than is commonly suggested in western circles, though not as much as claimed in the quotation.

Apart from direct pressures of a political nature, there are always incentives which, though politically inspired, act through economic channels to induce farm committees to vary rotations in particular directions. This is, of course, also true of western countries which experience government interference in agriculture. Some of the differential changes in prices for agricultural products in the Soviet Union have been mentioned in the sections on prices and marketing.

In the last resort, as anywhere, the farms must adhere to certain practices in their rotations if they are to fulfil their plans, make profits and conserve soil resources. Rotation practices, therefore, vary according to climatic conditions in different parts of the country, to soil variations and to the supply of inputs, notably fertilizers, as well as to the range of enterprises

[1] V. Shubin, *Pravda*, 16.9.66, 2; *C.D.S.P.* 18 (37), 24.
[2] An example is quoted on pp. 154–155.

on the farm. Shubin illustrates rotational practices advocated for the Volgograd area, where dry-steppe conditions prevail:

> We have many farms with large tracts of ploughland arranged on the basis of several divisions or integrated brigades. In these conditions, the combining of crop rotations justifies itself. In the vicinity of livestock sections, they set aside two or three fields for the crop rotation. On one of them they grow corn (maize) for silage or green fodder. The third field is planted to annual or perennial grasses for obtaining green forage, hay and hay meal. But on the remote six or seven fields of the basic crop-rotation tract they employ the fallow-grain type of alternation: one field of fallow, two fields of winter grain, one field of legume or groat crops and two or three fields of spring grain crops, primarily wheat. In addition to field-crop rotations it is necessary, in our view, also to introduce specialized fodder-crop rotations.[1]

The recurring crises in grain supplies resulted in the past in minimal rotations in grain areas as year after year reliance was placed on the richness of the chernozem soils and the drier climatic areas to produce the required crops. The poor results which were inevitable in climatically poor years were accentuated by bad husbandry and lack of fertilizers. Khrushchev's decision to create a new grain base in the east put at risk the soils of the virgin lands but it permitted a reduction in the amount of grain grown in the Ukraine and adjacent grain-growing areas. The area under all grains in the Ukraine was cut from 20 million hectares in 1950 to 13·8 million in 1960. The figure then became relatively stable at about 16 million hectares. Reductions in the areas devoted to grains were also made in the Belorussian, Baltic, Transcaucasian, Moldavian and Central Asian republics and in western and northern regions of the R.S.F.S.R.[2] In Kazakhstan, the grain area rose from 6 million hectares in 1950 to 24 million in 1965, falling back to 23·7 million in 1966 and 22·7 million in 1967. For the U.S.S.R. as a whole, the total area under grain rose from 103 million hectares in 1950 to 128 million in 1965, with a reduction to 124·8 million in 1966 and 122·2 million in 1967.

[1] V. Shubin, *Pravda*, 16.9.66, 2; *C.D.S.P.* 18 (37), 24.
[2] *N.kh. SSSR 1967*, 353; *N.kh. RSFSR 1967*, 183–184.

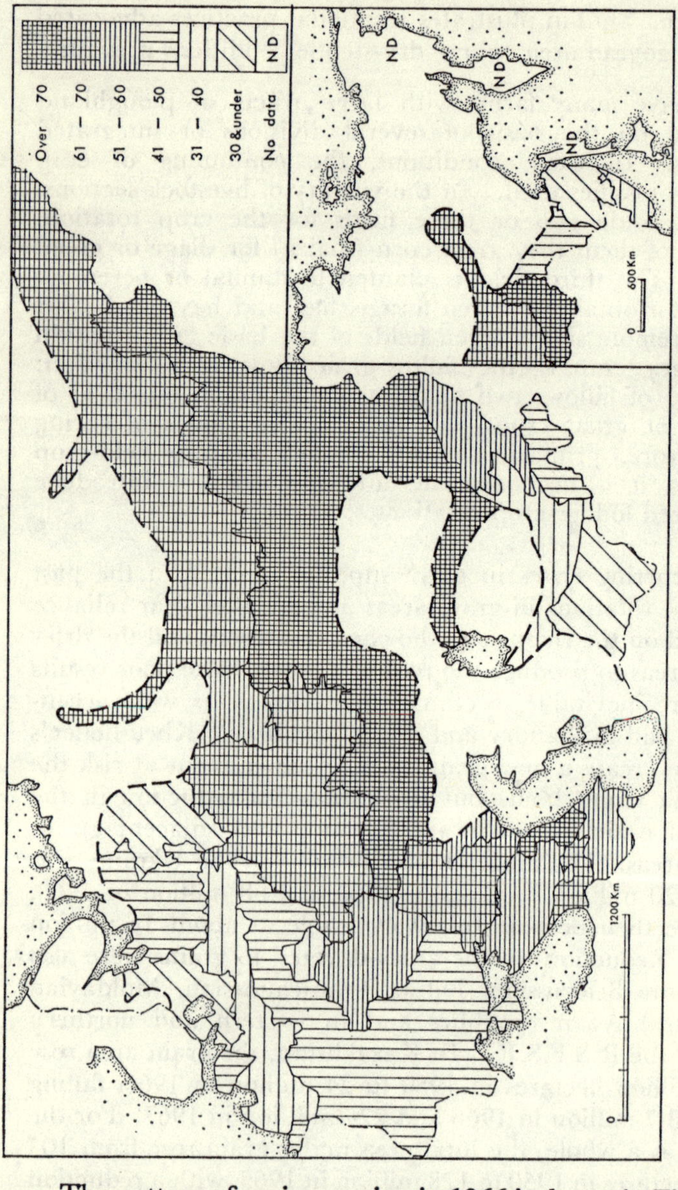

8. Percentage of sown area under grains, 1966

Specialization in grain growing to the extent of over 60 per cent. of the sown area (almost 80 per cent. in Kazakhstan) is shown in this map to have been applied to the steppe and wooded steppe areas from the Volga to central Siberia. The development of these eastern lands permitted a reduction in the grain area in the Ukraine and Belorussia from about two-thirds of the sown area in 1940 and 1950 to less than one-half in 1966

The pattern of grain growing in 1966 is shown in Figure 8. A fairly consistent gradient of increased concentration on grain is evident from north-west to south-east as far as the arid regions of mid-Kazakhstan and Central Asia, with the greatest intensity in the belt stretching eastward from Saratov and Orenburg oblasts into north Kazakhstan. In spite of some trimming back of grain growing and replacement by fodder crops in the former virgin lands the sown area there is still too much devoted to grain for good conservation practices to be possible and it is likely that there will be some swing back to a rather higher percentage of sown land under grain in the south-west.

Grain is not, of course, a homogenous commodity and it is preferable to turn to the individual grains for more detailed study.

WHEAT

Of the arable area, the greatest share is now devoted to spring wheat, which took the lead from rye (a winter-sown crop) some twenty years ago. In 1968, spring wheat occupied just over 48 million hectares, more than twice as much as winter wheat or barley (respectively 19 and 17 million hectares) which occupied second and third places. In the R.S.F.S.R., spring wheat occupied over 30 million hectares, barley about 10 million, rye 9 million and winter wheat 9 million. The spring wheat area is actually somewhat greater than these figures show, because the fallow land (about 17 million hectares for the U.S.S.R. as a whole) is predominantly part of the spring wheat rotation land.

The increase in the wheat area reflects the demand for wheat as a bread grain to replace rye consequent upon the rising standard of living. The shift in emphasis from winter to spring varieties, however, followed the development, especially from 1954 to 1958, of the virgin lands, mainly in the dry areas of western Siberia and north Kazakhstan, where spring-sown wheats are most appropriate. The increase in the wheat area in the east has lessened the pressure on the 'old' winter wheat lands of the Ukraine, north Caucasus and central black-earth areas, though these areas have actually increased their winter wheat acreages.

Winter wheat is the preferred crop, since autumn sowing permits rapid spring growth with maximum use of the moisture in the soil and the spring rainfall, but it cannot withstand very low soil temperatures, and is accordingly restricted to the more southerly areas where very hard frosts are not normally encountered.

Spring sowing in the eastern wheatlands is necessary because the snow cover is insufficient to protect autumn-sown grains, and because of the short period of relatively warm weather available for cultivation after the grain harvest. Some winter wheat is, however, grown in Siberia, mainly for experimental purposes. The success of spring sowing depends on adequate moisture being available and so, spring precipitation being uncertain, measures are taken to conserve soil moisture. These measures include fallowing and cultivation which does not turn over all the stubble of the previous crop as a mouldboard plough does. The stubble then helps to hold snow, preventing deep freezing and providing soil moisture.

Even the best spring varieties of wheat will not yield as well as good winter varieties sown at the right time. Furthermore, spring wheats generally occupy the less favourable lands, and so average yields of spring wheat are low. The best average yields for spring wheat in the U.S.S.R. were claimed for the 1966 and 1968 harvests—12 centners per hectare, compared with a range of 5·5 to 10 in most preceding years.[1] Winter wheats usually yield 14–17 centners per hectare, and the 1966 average was claimed to be 20·4.

Because of the total effect of the various factors, the regional variation in yield for both winter and spring varieties is as great as the annual variation of the whole country's average. The regional figures for the R.S.F.S.R. in 1965, a moderately good year for winter wheat, but a poor year in the main spring wheat areas, are compared in Table 13 with the figures for 1964, which showed, in general, the opposite trends.

Thus, in the R.S.F.S.R. in 1965, the average yield of spring wheat was under half that of winter wheat, whereas in 1964 it was over three-quarters that of winter wheat. In the North-west and Central regions in 1965, the spring wheat yields were above average, but 14 and 17 per cent. below those of winter wheat. In 1964 both regions showed lower averages for both, but in the Central region the spring wheat average equalled that of winter wheat. In the North Caucasus, in 1964, spring wheat almost equalled winter wheat in yield, but in 1965 it declined to

[1] *N.kh. SSSR 1968,* 351.

TABLE 13

AVERAGE WHEAT YIELDS IN THE R.S.F.S.R. ECONOMIC REGIONS
1964 & 1965
(centners per hectare)

	1964		*1965*	
	Winter wheat	*Spring wheat*	*Winter wheat*	*Spring wheat*
R.S.F.S.R.	12·9	9·9	14·1	6·8
North-west	9·7	5·9	10·4	8·6
Central	7·7	7·7	12·0	10·3
Volga-Vyatka	8·0	7·4	13·3	7·2
Central Chernozem	15·3	12·6	16·3	12·6
Volga	10·4	10·8	10·5	8·2
North Caucasus	14·3	13·9	15·1	7·9
Ural	7·1	9·9	10·2	6·3
West Siberia	—	9·9	—	4·8
East Siberia	—	9·1	—	8·9
Far East	—	7·5	—	7·2

Source: N.kh. RSFSR 1965, 247.

half the winter wheat yield, and yields were lower throughout the east, so lowering the R.S.F.S.R. average. In the relatively stable Central Chernozem region, both winter and spring wheats averaged higher yields for the two years than in any other region, and the relationship between the two varieties, or, rather, groups of varieties, was about normal.

In 1964 the Siberian averages for spring wheat compared favourably with those of several other areas, but in 1965 in west Siberia the yields dropped to 4·8 centners per hectare—nearly as low as in the drought season of 1963. Then, West Siberia's average was only 3·6 centners per hectare, half the R.S.F.S.R. average, itself one of the poorest figures on record. Between these two years, however, the average was 9·9. The highest annual average for spring wheat in the two years was that of the North Caucasus. Here, too, fluctuation in spring wheat yields is considerable. Winter wheat, however, is more reliable, and often yields more highly in the North Caucasus region than in the Central Chernozem region, as in 1966 and 1967.[1]

[1] *N.kh. RSFSR 1967*, 231.

The regional variation in yields goes far to justify the controversial virgin lands programme. Relying on the grainlands of European Russia and the Ukraine would clearly have left the Soviet Union worse off in wheat supply than it now is with the productive capacity of the Siberian and Kazakh lands increasing supply. Further, the grain growing of the eastern lands has facilitated diversification in the western areas, especially into the fodder crops essential to support the increase in livestock production needed to supply dairy and meat products to the western cities.

More balanced production, too, reduces the burden on the transport network. If the Ukraine and North Caucasus areas had maintained their former large surplus in wheat production, a growing amount would have had to be moved to the expanding wheat-deficit areas in more northerly and easterly areas, while livestock products or fodder were sent into the Ukraine and North Caucasus. Now, southern wheat production approximately balances needs, and the wheat-deficit areas of the north are supplied direct from the eastern areas of surplus. It is not always as simple as this, but at least the balance has been improved and the food supply better secured by the eastward expansion of wheat growing.

The cultivation of wheat at the northern and eastern limits of its climatic tolerance is more relevant to the feeding of local populations and the reduction of long hauls, particularly to supply mining, forestry and other remote communities, than to the increasing of total production. Improved, hardy varieties of wheat are grown at least as far north as 60 degrees and as their frost resistance and speed of ripening (some now take only 60 days from sowing to maturity) are further improved the limits can be pushed back into the colder zones. Vernalization, i.e., exposure of the seed to abnormal controlled temperatures, and preplanting germination are among the techniques employed for these northern areas and in Transbaykalia and the Far East.

The contribution of different regions to the increases in wheat output up to 1967 (compared with 1940) are shown in Figure 9. The role of the former virgin lands needs no further comment at this point but it is notable that the main increases in production came otherwise from traditional grain-growing areas—the Ukraine, north Caucasus, central black-earth and other western areas, with only slight increases or decreases recorded in the Baltic republics and northern oblasts of the R.S.F.S.R.

Comparison of yields over a long period faces problems of lack of comparability in data. It appears, however, that whereas

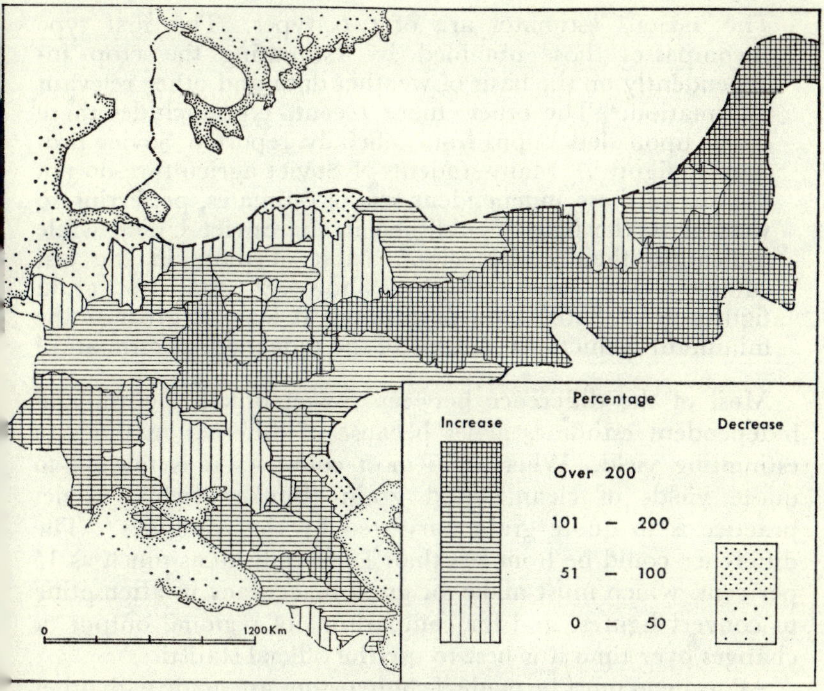

9. Wheat: percentage change in production, 1940–67

The changes in the main area of wheat production are shown in this map. Comparison with Figure 8 shows the expected correlation between the high percentage of the area under grain and the large increases in production in the case of the virgin lands, not adequately reflected in the limited categories of the map. In the Volga lands and the western Ukraine, however, the more moderate increase in production has occurred with much lower proportions of land under grain. This results from a combination of reduction in some grains, especially rye and low-yielding spring wheat, and improved yields of winter wheat

between 1913 and 1940 there was little improvement in average yields of wheat, in recent years there has been marked improvement. This is most noticeable in winter wheat, as would be expected from its restriction to the better land.

Spring wheat, however, despite its large-scale expansion on to poorer areas, also shows improved yields.

Walters and Judy have summarized the attitudes of western students of Soviet agriculture to the grain statistics published in the Soviet Union as follows:[1]

> The various estimates are of two types. The first type encompasses those obtained by estimating the crop independently on the basis of weather data and other relevant information.[2] The other, more recent, type includes those based upon deductions from officially reported Soviet production figures.[3] Many students of Soviet agriculture do not use any of these independent sets of estimates, preferring to work with the figures published in the Soviet Union while indicating their deficiencies.[4] Most of those in the last category have made it clear, however, that the grain figures are inflated by a percentage at least as great as the minimum deductions in some of the independent estimates.[5]

Most of the difference between Soviet official figures and independent estimates arises because of different methods of estimating yield. Whereas in most countries it is normal to quote yields of clean, dried grain (bunker yield), Soviet practice is to quote grain harvested by the combines.[6] The difference could be from less than 1 per cent. to as much as 15 per cent. which must make for great inaccuracy in attempting to convert figures, and for comparisons of regional output or changes over time it is best to use the official statistics.

Allowances must be made if comparisons are made with other countries but such comparisons are not very rewarding. Clearly, to compare the average of the whole of the U.S.S.R. with that

[1] Walters and Judy (1967), 311–312.

[2] Walters and Judy here note that these include the estimates of the U.S. Dept. of Agriculture in *The 1964 Eastern Europe Agricultural Situation*, Washington, 1964, and of the Central Intelligence Agency, *Production of grain in the USSR*, Washington, 1964.

[3] Including those of Kahan (1963) and Klatt (1964).

[4] This is the position taken in this book. While there is no advantage to the academic researcher in quoting statistics inflated for political reasons, there is equally no advantage in quoting revisions of these figures which cannot be proved, some of which also may not be without political inspiration.

[5] Walters and Judy here note that most have indicated a range of from 10 to 15 per cent.

[6] *N.kh. SSSR 1964*, 814; *1967*, 923, etc. Walters and Judy (1967) examine the matter in some detail. The very inaccurate 'biological' method of estimating yields from the standing crop was abandoned in 1956.

for the U.S.A. for any crop is meaningless, since both countries encompass great internal variation, quite apart from the fact that the U.S.S.R. lies in latitudes almost wholly north of the U.S.A. Parts of Canada offer more relevant comparisons, but at best such comparisons offer no more than a yardstick which may indicate the level of achievements.[1] Thus, at a conversion rate of 1·48 bushels per acre (one centner per hectare), a yield of 16 centners per hectare approximates to 24 bushels per acre, so that even this relatively high Soviet average is still far less than is obtained in most countries. The question of whether the Soviet authorities inflate grain yields for political purposes remains unsolved but Soviet statistics showed wheat production from 1963 to 1965 inclusive as only 183·8 million tons, an annual average of about 60 million tons, which does not suggest distortion. The 1966 figure was, however, 100·5 million tons, and such a great improvement as this total must represent, even if not completely accurate, would surely not require inflation for political reasons. If it were inflated deliberately, then the drop to 77·4 million tons in the following year, when the harvest was poorer, would have been exaggerated, or, if deliberately inflated figures were maintained, this practice would have to be continued or the credit for subsequent improvement again lost.

The three-year average, 1964–66, at 78·2 million tons, was some 17 million tons or 28 per cent. better than the preceding three-year average. Further, it included one good, one bad and one medium year, and was reasonably representative. From the official statistics without modification, in the early 1950s, as in 1940, wheat production per head of the (enlarged) population was no higher than in 1913. In 1950 total wheat production was put at 31·1 million tons, in 1966 at just over 100 million tons. Even the three-year average of 1964–66, 78·2 million tons, would have given the increased population over twice as much wheat per head as in 1950.

The total yield of wheat has continued to fluctuate, as may be expected, but within narrower limits than in the past, presumably because of the geographical spread of production between the older and newer areas. The total reported harvest for 1968 was 93·4 million tons and for 1969, 79·9 million tons.[2]

[1] See Zoerb (1964) and Johnson (1963).
[2] *N.kh. SSSR 1969*, 326.

Of course, the wheat crop is not entirely used for human consumption, and production of rye has decreased by two or three million tons a year, but there is certainly a vastly better supply of bread grains available now in the Soviet Union than at any time in the past. The expansion on to the virgin lands and improved yields generally have made it possible to reach this position, while yet diversifying production on land which formerly specialized in wheat production. It has also made it possible for the Soviet Union to return to its traditional role of a net grain exporter, forsaken during the years 1964–66 when grain imports amounted to 21·4 million tons compared with exports of 11·4 million tons.[1]

RYE

Rye, the traditional bread grain of the forest regions, now occupies much less land than even a decade ago. In 1968, it reached an all-time low of 12·3 million hectares compared with 16·2 million in 1960, 23·6 million in 1950, 23·1 million in 1940 and 28·2 million in 1913. Rising standards of living and the greater availability of wheat flour, as well as the hardier strains of wheat, have contributed to this reduction. However, the higher prices that have encouraged wheat production apply also, though rather less, to rye, and the crop remains valuable in areas beyond the limits of reliable wheat growing. Though rye is almost entirely winter-sown, spring sowing enables its limits to expand in the northern regions and eastern Siberia. The main rye areas are, however, still in Belorussia, extending into the northern Ukraine and in the Volga country east of Moscow, where soils are leached and acid, and rye occupies over 20 per cent. of the sown area.

Yields of rye have not risen significantly. Though returned as averaging 10·5 centners per hectare in 1967, this was little better than the yields of pre-war years.[2] There seems, on balance, little likelihood that the production of recent years, averaging 13–16 million tons,[3] will be raised in proportion to grain crops as a whole, though rye may yet receive renewed attention in order to produce more grain in the northern areas.

[1] *N.kh. SSSR 1965*, 671–672; *1968*, 659–660.
[2] *N.kh. SSSR 1967*, 378.
[3] *N.kh. SSSR 1967*, 377.

BARLEY

The next most extensive crop, comparable in area to winter wheat, both in the U.S.S.R. as a whole and in the R.S.F.S.R., is barley, which, though classed as a spring grain, may also be autumn-sown. As in many other countries, there has been a dramatic increase in the use of barley in the U.S.S.R., over 17 million hectares annually from 1963 to 1968, compared with 11 million in 1960, 8 million in 1950, 10·5 million in 1940 and 13 million in 1913. The R.S.F.S.R. shows a similar trend but now provides about half the total barley area as compared with about one-third in 1913. In general, however, the great expansion in the wheat area has lowered the share of barley in the total grain trade.

The great advantage of barley is its tolerance. Barley begins its growth at lower temperatures than other cereals, is less sensitive to frost and drought and matures more quickly. Strains have been evolved suitable for the most northerly cropped lands, well north of 60 degrees, for dry areas in the south and for mountain lands. Cultivation of barley is accordingly extremely widespread. It is particularly found throughout the main arable belt from the Ukraine, north Caucasus and central black-earth regions to west Siberia. It has been included in sowings in the virgin lands, while the extreme southern areas of the U.S.S.R. (south Ukraine, Caucasus, Transcaucasus and Central Asian areas) are those where the main autumn-sown barley crop occurs.

Barley shows well the increases in yield attained by improved seed, fertilizers and techniques in recent years. Official averages gave 14·1 centners per hectare for spring barley in 1966, and 12·7 in 1967. Yields before 1960 were generally below 10 centners per hectare. For winter barley, the improvement is much greater; from pre-war figures little better than for spring barley to a U.S.S.R. average of 17·2 centners per hectare in 1966, while the R.S.F.S.R. claimed an average of 20·7 for its winter barley in 1965.

Expanded areas and improved yields have doubled the output of barley as compared with 1940 and even quadrupled it as compared with the bad year of 1950, the 1965–67 average being reported as 24·3 million tons. The grain is mainly for fodder but some malting varieties are grown.

OATS

The oats area has declined in much the same proportion as the rye area, the diminishing number of horses being one reason. From 20·2 million hectares in 1940, the area declined fairly steadily to 6·6 million in 1965, but then recovered to 9 million hectares in 1968. The reduction having been greater in southern regions, where maize offered greater possibilities than in the north, the share of the R.S.F.S.R. has increased somewhat from the 75 per cent. of the oats area formerly typical. The crop is widely cultivated, but mainly between about 50 degrees and 60 degrees N., with particular popularity in the central, central black-earth and Volga regions. Moisture requirements are higher than for barley and frost tolerance rather less, but, like barley, it is hardy and can be sown in the autumn, although usually sown in the spring. It also tolerates acid soils better than barley, hence its wide use in the forest zones, where neither excessive frosts nor serious droughts threaten it.

Yields have improved moderately, from reported national averages of 8·9 centners per hectare in 1913 and only 8·1 in 1950 to 11·8 for the three years 1965–67. With good yields and higher sown areas, the 1967 production was reported as 11·6 million tons, compared with 9·2 million in 1966 and 6·2 million in 1965. The 1940 and 1913 figures were about 17 million tons.

OTHER SMALL GRAINS

Millet has some importance in areas particularly liable in the growing season to drought but not to hard frosts. It grows quickly, once temperatures exceed 10–12 degrees C., in regions of relatively short summer days, and makes full use of the irregular summer rainfall of the steppes. It is, then, suited to the summer rainfall areas, but requires irrigation in dry regions of winter and spring rainfall, such as the eastern Caucasus and central Asia. The chief millet areas are the north Caucasus, lower Volga and southern Ural steppes.

The area of millet has not varied greatly, averaging 3·4 million hectares (1965–69), of which rather under two-thirds was in the R.S.F.S.R. In 1913 the area was little larger, but significantly so before and during the Second World War. The 1967 reported average yield of 6·8 centners per hectare was higher than usual, though 10·6 was claimed as the R.S.F.S.R. average in 1964.

Buckwheat and rice constitute minor but valuable food crops. Buckwheat grows well in the forest-steppe areas and the southern parts of the mixed forests, having a shorter growing period than millet (70–85 days), though sensitive to frost and drought. It is important in Baykalia and the Far East as well as in European Russia. The average yield of 5·7 centners per hectare for the 1965–67 period gave an average total production of just over one million tons for the U.S.S.R., about 10 per cent. less than in 1913, but from nearly a quarter less sown area.

For the whole U.S.S.R. the buckwheat area increased following the 1965 upward revision of prices, from 1·4 million hectares in 1964 to 1·8 million in 1966, but this was less than the 1940 figure and it diminished again slightly in 1967 and 1968, so the long-term trend appears still to be downward.

A feature of buckwheat is its high yield of nectar, with which is coupled strong dependence on pollination by bees, and so its cultivation is often linked with bee-keeping. The grain is used for both fodder and porridge.

Though never a major crop in the Soviet Union, because of climatic limitations, rice has received increasing attention this decade. Improved yields have been mainly responsible for increased output from about 200,000 tons annually before 1960 to 580,000 tons in 1965 and over one million tons in 1968, the average yield reaching 34·1 centners per hectare in that year.

It is grown mainly in Central Asia and the Caucasian regions, while the R.S.F.S.R. contribution comes mainly from the Far East, especially the Maritime kray, near Lake Khanka. In Central Asia the main areas are the middle reaches of the Syr Darya, the Tashkent area, the Fergana basin and the lower Amu Darya; in the Caucasus region, the Kuban delta, the middle Kura River valley and the Lenkoran lowland are most important. All these areas have developed irrigation systems, and, as these are extended, rice growing may well be expanded. There is considerable demand for rice in the big cities, as well as in the growing regions.

The Kuban Rice Experimental Station has developed new

varieties, with emphasis on spreading the dates of maturity of crops for economical harvesting. Cultivation in autumn and sowing in spring is favoured, and has established the high value of perennial grasses preceding rice in rotations. This succession improves response to mineral fertilizers.[1]

Saline soils in Rostov oblast and in the lower Terek River (north Caucasus) are considered to justify development of irrigation for rice growing.[2] Existing irrigation systems have not, however, been used to the best advantage and recent work by the Kazakh Research Institute has been directed at improving water utilization. It recommends deep flooding but has proved that shallow flooding may be used at times of water shortage, providing weeds are rigidly controlled.[3]

MAIZE

Nothing in the Soviet cropping programmes in recent years has attracted as much attention in the west as the development of maize cultivation under Khrushchev. Impressed with the value of maize as a fodder crop, particularly after his visit to the United States, Khrushchev praised its merits excessively and promoted a rapid extension of the maize acreage. It is, however, an exaggeration to state that he 'insisted on corn being grown practically throughout the vast land.'[4] Maize did displace other crops and grasses to a considerable extent, but the disadvantages of the expansion of grassland under the Vilyams system had been realized, and oats and rye, the areas of which were much reduced to make way for maize, also appear no longer to justify their earlier importance in the cropping pattern. Khrushchev almost certainly over-emphasized the possibilities maize offered in Soviet conditions, and there has since been a reduction in the maize area, but, strikingly, not to its pre-1960 area, nor have rye and oats been permitted to take up the areas so freed. As noted above, it is wheat and barley that have been given further impetus, as might be expected from Soviet climatic ranges and the relative demand for the different grains.

[1] Brezhnev and Minkevich (1958), 1961, 39–40.
[2] Brezhnev and Minkevich (1958), 1961, 40.
[3] Goryunov, Petrunin and Sirgelbayev (1967).
[4] Volin (1967), 14.

In the Soviet Union, maize is used as:

(1) Ripe grain
(2) Ripe silage
(3) Green silage
(4) Green fodder

Introduced into the Carpathian and Caucasian borderlands in the sixteenth and early seventeenth centuries, it spread into the Ukraine and Kuban lowlands in the nineteenth century and received stimulus from V. V. Talonov and others early this century. Between 1927 and 1933 the maize area increased to over double that of the pre-Revolution years as agriculture was stimulated by the First Five Year Plan. It reached 12 per cent. of the sown area in the north Caucasus, compared with 3 per cent. in 1913. There was a marked fall between 1933 and 1940, mainly where it was grown for fodder. Between 1940 and 1950, in the Ukraine, Khrushchev, as Party secretary, was already urging its wider use, and there its area expanded by nearly a million hectares during this period. Elsewhere in the U.S.S.R., it declined, and decline also occurred in the Ukraine after Khrushchev left its administration.[1]

Khrushchev began his general campaign to increase the use of maize in his address to the September 1953 Plenum and when, that winter, autumn-sown grains were badly hit by severe frosts, the opportunity was taken to replace some of the small grains by maize. In spite of production setbacks the maize area reached successive new peaks in 1955, 1956, 1960 and 1962. From 1955 onwards, the area of maize for grain was exceeded by that for combined ripe and green silage uses, and from 1956 to 1963, except in 1961, it was also exceeded by maize for use as green fodder. The silage and green fodder totals fell slightly in 1963, and reduction in all maize areas in 1964 brought the figures back to 1960–61 levels. A further decline occurred in 1965 nearly to the 1959 level, since when there has been comparative stability. At 22·5–23·5 million hectares, the 1965–69 areas are rather less than the 1956 figure, so that corn has recently been allocated about the same area as

[1] Anderson (1967a) provides the basis for this summary of maize growing. He quotes numerous authorities.

in the earlier phase of expansion under Khrushchev, many times what it was before that phase, and about 62 per cent. of its 1962 peak.

Maize for grain has declined more than the silage crops in proportion to the total, occupying in 1965–68 between 3 and 3·5 million hectares, rather less than half the peak figures. Climatic conditions limit ripening to the southern areas. The Ukraine is dominant here, with the north Caucasus and Don-Volga lowlands the most important in the R.S.F.S.R. Most of the rest of the R.S.F.S.R. has under the 2,400 day-degrees of accumulated temperature above the threshold of 10 degrees C. regarded as necessary for ripening maize for harvesting as grain. The irrigated valleys of Central Asia contain the only other significant sowings of maize for grain.

As the R.S.F.S.R. areas are climatically inferior to the Ukraine for ripe maize, it is not surprising that the maize expansion programme did not emphasize ripe grain as much as in the Ukraine. The peak in the R.S.F.S.R. came in 1961 at 2·3 million hectares, somewhat under one-third of the total U.S.S.R. area for grain maize, and the contraction was more severe, to a 1965–66 average of 700,000 hectares, only 22 per cent. of the U.S.S.R. figure.[1]

Silage and fodder sowings of maize in the R.S.F.S.R. reached a peak in 1962, when the U.S.S.R. figures also reached their highest, the Russian republic figure being 19·1 million hectares out of the U.S.S.R. total of 30·2 million hectares, i.e., 63 per cent. The decline has been rather steeper than in the U.S.S.R. as a whole, the 1966–67 average in the R.S.F.S.R. of 9·7 million hectares being one-half of the U.S.S.R. totals.[2]

Although sowings for grain and for silage or green fodder crops are separately distinguished in Soviet agricultural statistics, the grain sowings being classed as cereals and the others as fodder crops, there is, necessarily, some overlap. In particular, maize sown with the hope of yielding ripe grain often has to be harvested for silage because it has failed to mature. Whereas maize can regularly be harvested fully ripe in the southern areas, as indicated above, it is planted specifically for silage and green fodder over wide areas of Russia and Siberia south

[1] *N.kh. SSSR 1967*, 359.
[2] *N.kh. SSSR 1967*, 348; *N.kh. RSFSR 1967*, 175.

of about 55 degrees latitude. East of the Ob its occurrence is sporadic in valleys of the Yenisey, Angara and Amur.

The widespread adoption of maize as a fodder crop has been important in the expansion of livestock rearing. In addition, it is valuable in a rotation, helping weed control and leaving, after harvest, a stubble and root system which helps conserve snow and soil moisture. It is therefore favoured before spring wheat in the eastern wheatlands, even though the climate is too dry to be near the optimum for maize. Maize makes little growth before the temperature reaches 10–12 degrees C. and hence is essentially a crop of summer growth. Short days do not hinder it, relatively, but it cannot withstand even light frosts. Fast-growing varieties mature in three months but even these cannot be harvested fully ripe in the eastern grainlands.

Undoubtedly, it is the flexibility, as well as the prospect of high yields of fodder units, that has encouraged the Soviet authorities to persist with maize. If the cob can be utilized for grain, the stalk can be ensiled, or the whole can be ensiled at the milk-wax stage of ripeness, or the plant can be used for green fodder. If the season is insufficiently moist for the best results, the deep rooting habit of the plant will nevertheless enable it to draw on reserves of moisture in the soil.

The reported yields for grain were higher for the five years 1965–69 than ever before, the U.S.S.R. averages being about 26 centners per hectare, with 28·6 claimed for 1969.[1] This suggests that concentrating the grain crop in the more suitable areas has been beneficial.

Grain maize is the highest yielding grain in nearly all regions in which it is grown, excelled only by winter wheat and barley along the arid margins of its cultivation in the southern Ukraine and northern Caucasus. Its returns are aided by the low seeding rate of less than 0·5 centners per hectare against 1·5–2·0 centners for small grains.[2] Grain maize, it has been estimated, probably accounted for nearly 20 per cent. of the concentrates fed to Soviet livestock by 1962 compared with about 7 per cent. of a much smaller total in 1953.[3]

Ripe maize silage is made as far north as the 2,200 day-degree zone of accumulated temperatures, i.e., from mid-Belorussia

[1] *N.kh. SSSR 1967*, 380; *1969*, 328

[2] Anderson (1967a), 118–119.

[3] Anderson (1967a), 119.

to the Kazakh-Siberian borderlands. It was reduced in the Ural and Volga areas in 1962 because of the effect of frosts and droughts, but was extended in the north Caucasus and Ukraine. This expansion occurred partly on land formerly fallowed before the sowing of winter grains in order to obtain a silage crop before autumn sowing. The improved weed control and increased fodder production were considered to offset any depletion of soil moisture, but this practice has been blamed for the loss of winter wheat crops.

Maize for ensiling in the unripe stage and maize for fodder have not been separated in recent Soviet statistics. Anderson examined the distributions of the two forms between 1953 and 1962,[1] and noted that in 1959 there were two distinct patterns. First, in the steppe Ukraine and north Caucasus maize was sown after harvests of winter grain to utilize residual moisture and thermal reserves and on black fallow. Secondly, in a broad belt from the Baltic to Lake Baykal, it appeared to account for all the maize that could be harvested with available labour and machinery. Silage may have been the object of much of the maize which was, in fact, used as green fodder because of shortages of labour and machinery at critical times. This would be specially so when the orders to plant maize reached the excesses of 1962.

In the Baltic republics and the mixed-forest zones of Russia and Siberia, where accumulated temperatures exceed 1,800 day-degrees, maize can be grown with reasonable expectation that it will provide good silage. Early-maturing varieties should reach the milk-wax stage of ripeness. Clearly, however, for maximum benefits, there has to be a combination of suitable varieties, adequate fertilizers and the machinery and labour needed for handling the crop.

Whatever the excesses of the maize programme before 1963, there seems little reason for doubting that Khrushchev was right in advocating its cultivation over a much wider area than previously. Maize silage has provided much of the succulent fodder fed to dairy cows in winter, which has largely enabled milk production to be increased markedly, and grain maize has also improved livestock productivity. Some losses have occurred from diversion of effort from other crops, but maize products

[1] Anderson (1963), 260–262; (1967a), 121–124.

should, in general, have contributed more than the hay and oats crops they have displaced. Harvesting in the green stage has been more wasteful and here the limitation of recent maize sowings may prove beneficial, and be the key to a reasonably satisfactory pattern of maize growing in the U.S.S.R.

GRAIN LEGUMES

Soviet statistics class the grain legumes as *zernobobovyye.* Various kinds of beans and peas are the main crops in this category. The combination of a yield of protein-rich fodder and enrichment of the soil through fixation of nitrogen makes the legumes attractive crops. Together with maize and sugar beet, they received official blessing at the March Plenum in 1962 as intensive fodder crops which should, as far as possible, replace low intensity crops, especially hay and oats. The expansion achieved in sowings of the legumes followed that of maize and by 1963 the area sown with them (10·8 million hectares) had more than trebled since 1960. As with maize there was then a decline—to 6·8 million in 1965 and 5·1 million in 1968.

They are widely grown, with some concentration in the central black-earth region and Volga area, the Ukraine, Moldavia and Transcaucasia. In the southern European parts kidney beans are most common. Lentils are favoured in the wooded steppe zone in the Volga, central black-earth and Ukraine 'right bank' areas, while peas are commoner in the north.

Low yields were experienced in the early years of the drive to increase the grain legume area, no doubt partly because the farms were inexperienced in handling them. In 1968, however, yields attained a reported average for the whole Union of over 14 centners per hectare, which should make them more competitive with other grains, particularly in view of the high protein content of the stems and leaves as well as of the grain. At about 7 million tons, the harvest is well over twice the average before the expansion.

FALLOW

Before considering the production of industrial and fodder crops we may conveniently consider the place of fallowing in the Soviet systems of agriculture, particularly as fallow is usually associated with grain cropping. Land which is lying fallow is not directly productive, and in that sense is negative, but it is aimed at accumulating moisture and nutrient before the sowing of a crop, and therefore has a positive function. In an economy where maximizing production per hectare is important, fallow is to be avoided where it is not essential to conservation or the success of the following crop.

Without adequate fertilizers or restorative crops and moisture, fallowing is required, and in old Russia, as in other European countries before the introduction of root crops and leys, fallow was essential in the primitive rotational system. Even in relatively moist areas it was valuable, as it facilitated the destruction of weeds and the restoration of fertility, but in dry areas it was considered obligatory. Selective weed killers and increasing availability of chemical fertilizers, together with root crops, have greatly reduced the need for fallow in the moister lands, though, with certain modifications, it remains important in areas dry enough to necessitate dry-farming practices.

Fallowing was attacked by Khrushchev as wasteful use of land and was markedly reduced during his supremacy. From 28·9 million hectares in 1940 and 32 million in 1950, it was cut down to 17·4 million hectares in 1960 and 6·3 million in 1963. That the elimination of fallowing in some areas resulted in severe damage to the land and diminished yields of crops was apparent to many observers, both Soviet and foreign, and after Khrushchev's fall a rise occurred in the area of fallow to 11·6 million hectares in 1965, 16·8 million in 1966 and 18·2 million in 1968.

Khrushchev was right, however, in criticizing reliance on fallow in some areas. Anderson pointed out that in 1959 fallowing appeared to be most predominant in the cool regions of pioneering agriculture, such as eastern Siberia, and in European Russia, as a remnant of peasant practice.[1] In these areas 12–17 per cent. of the arable land under cultivation was fallowed, where fertilizing and row-crop cultivation should have controlled weeds and raised yields. In Tselinnyy kray, however, under 10 per cent. of the arable land appeared to be fallowed.[2] Thus, the practice could, with advantage, be reduced in the cool northern and peripheral regions, but needed if anything, to be increased in the dry regions.

[1] Anderson (1963), 255.
[2] Anderson (1963), 257.

Clean fallowing is most needed in the virgin lands, where rainfall and snowfall together provide barely adequate moisture for crops, and perennial weeds are a serious problem. The campaign against fallow, based on the efficiency of weedkillers and intertillage, has been officially favoured but the advocates of fallow have obtained a more balanced approach to its use. *Pravda* carried in 1967 a forthright statement by the director of the All-Union Grain Farming Research Institute: 'Clean fallow is the best forerunner for spring wheat in any year, and is altogether indispensable in dry years . . .'[1] The Institute reported that clean fallow and mouldboardless tillage had given an average harvest over six years of 12·9 centners of grain per hectare, twice the general average for the area.

Referring to the arguments of the 'herbicide school', however, the director of the Institute agreed that fallow was less necessary in the Kuban region. The mild and moist period of some three months after harvest gave the soil time to accumulate moisture and nutriments and allowed weeds to be destroyed, whereas in the virgin lands, with perhaps only a few days available for this, clean fallow was necessary.

Earlier, Shubin had stressed the need for fallowing in the Volgograd and Saratov oblasts.[2] He claimed that the unjustified elimination of clean fallowing had adversely affected the production of the areas where winter wheat had formerly been grown mainly on clean fallow. Yields had been higher than for winter rye, spring wheat or barley, but had fallen when the winter wheat was sown on occupied fallow. In many cases areas sown to winter wheat failed and had to be resown to spring crops.

On the other hand, on an experimental sovkhoz near Kharkov, visited by the writer in 1968,[3] clean fallowing was considered unnecessary in this area provided adequate amounts of fertilizer and correct rotations were employed. The condition of the land and output appeared to support this view. This, however, was land classed as wooded steppe and there was no contention that fallowing was unnecessary on unirrigated lands further south.

[1] A. Barayev, *Pravda*, 16.2.67, 2; *C.D.S.P.* 19 (7), 27–28.
[2] V. Shubin, *Pravda*, 16.9.68, 2; *C.D.S.P.* 18 (37), 24.
[3] Described in Chapter 6.

CHAPTER 8

Industrial Crops, Fruit, Vegetables and Grass

An important part of the arable area of the Soviet Union—over 10 per cent.—is used for sugar beet, oil-bearing crops, fibres, tobacco and tea, the heterogeneous group known as industrial or technical crops. The common feature is that all are produced primarily for factory processing for food or raw materials, though in part they may be used for animal feed. From only 4·1 per cent. of the sown area in 1913, these crops were expanded faster than the sown area to comprise nearly 8 per cent. in 1940, since when the industrial crops section has roughly paralleled the further expansion of the arable area. About half the total area of these crops is in the R.S.F.S.R., but the distribution of the individual crops varies greatly, both within the Russian and within each of the other republics.

SUGAR BEET

Sugar beet is the principal source of sugar in the Soviet Union, though some cane sugar is imported, mainly from Cuba. Sugar beet has been cultivated in the Russian lands since the early nineteenth century, and has been greatly expanded since 1950. The area under this crop in the U.S.S.R. was then 1·3 million hectares (0·68 million hectares in the same area in 1913[1]). The need to expand sugar production to supply the increasing population at improving standards of living led to much increased sowings. A peak of 4·11 million hectares was reached in 1964, falling gradually back to 3·38 million hectares in 1969. The R.S.F.S.R. share of this total has been consistently about 40 per cent., much of it in the Central Chernozem region and nearby areas which are, in effect, an extension of the main zone of concentration

[1] *Strana Sovetov za 50 let* (1967), 128.

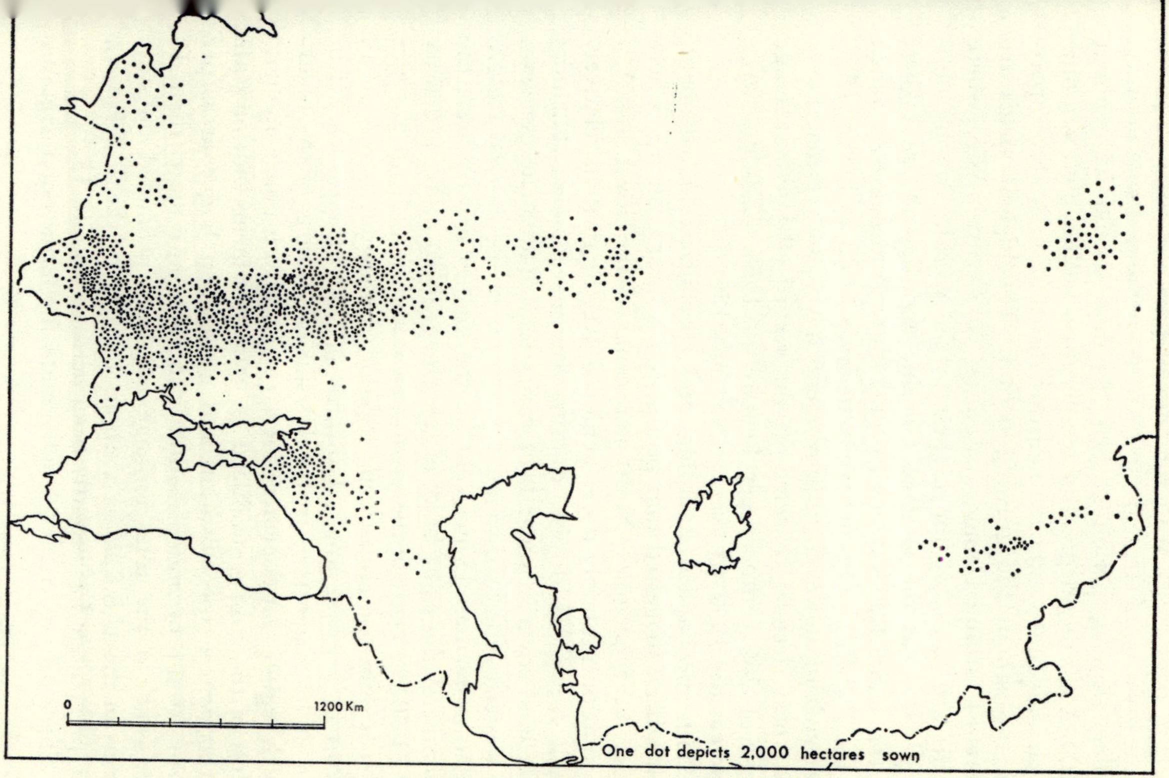

10. Sugar beet: distribution of cultivation, 1964

Since the Second World War the growing of sugar beet has been greatly intensified and the crop has been introduced into areas beyond its traditional zone of the Ukraine and south-west Russia. The Volga, Baltic, north Caucasus, Central Asian and western Siberian areas are outstanding on this map, but the south-west is still the main region for the crop. Source: *Atlas razvitiya khozyaystva i kul'tury SSSR*, 66.

in the Ukraine (Figure 10). All republics grow some sugar beet, in conformity with the present policy of some dispersal of the growing of strategic and widely demanded commodities.

The growing period of 125–135 days should have about 2,200–2,500 day-degrees of temperature with a high sunshine level. Adequate and even moisture distribution is also important throughout the growing period. The highest yields are therefore attained under irrigation in Central Asia, where yields are probably the highest in the world. Among non-irrigated areas the western Ukraine and Kuban are highest in the U.S.S.R. Soil acidity and lower sunshine over much of the R.S.F.S.R. non-chernozem areas result in lower yields. Everywhere, however, improvements have been registered over the past decade, reflected in the record all-Union average yield of 266 centners per hectare in 1968, (230 in 1967), whereas the 1959–63 average was only 156.

With mechanical handling and specialized management, it has been suggested that production is most economic where sugar beet occupies 20–30 per cent. of the sown area,[1] a proportion probably never achieved over more than relatively small areas. Hultquist, quoting various sources, concludes that land sown in sugar beet varies from under one per cent. in western Siberia through 3 per cent. in the eastern regions to no more than 15 per cent. in the highest density area, the country of the right bank of the Dnepr.[2] To attain a density of 20–30 per cent. of the sown area would be difficult as sugar beet is only one crop in the rotation, and is not grown on all farms in a given area. Upwards of 200 hectares per farm are under sugar beet in the Ukraine and parts of the black-earth region, but elsewhere the area per farm is much smaller.[3] Most of the crop is produced on collective farms, not on state farms, where specialization and consequent larger areas per farm would be more feasible. In fact, not more than 10 per cent. of the total tonnage has been produced on state farms in the U.S.S.R. as a whole, though for the R.S.F.S.R., the percentage has recently been between 14 and 18.[4]

There are, of course, good reasons for farms not increasing

[1] Nikitin and others (1966), 171.
[2] Hultquist (1967), 144.
[3] Hultquist, (1967), 147.
[4] *N.kh. SSSR 1968*, 321; *N.kh. RSFSR 1968*, 191.

their sugar beet area to that which would give the lowest cost per ton of beet to the mill, or even to the farm. A major problem, which increases with the size of the planted area, is the seasonal demand for labour. Even with the fullest mechanization, the crop needs 30–40 man-days per hectare, and more usually 40–60 man-days. This, however, should be compared with 180–200 man-days for non-mechanized working. Excessive sowings of sugar beet in relation to the supply of labour on the farms has necessitated seasonal recruitment of non-farm labour to the detriment of other work in many regions year after year.

The combination of high density of rural population with good growing conditions explains the early development in the Ukraine and in Kursk and Voronezh oblasts, with the later extension into Moldavia. Here, the wooded steppe was exploited, while eastwards in the same zone expansion has been more recent, and extends to the Bashkir A.S.S.R. Further south, the moister areas of Krasnodar kray are an important extension of the sugar-beet zone. Northward, the better soils of Belorussia, Latvia and Lithuania are also used. Eastward, Altay kray has considerable beet areas, Transbaykalia and the Far East have smaller developments, but severe labour problems in these areas prevent large scale operations.

While sugar production has been the main stimulus for growing beet, its fodder value is increasingly important. The beet-tops, foliage and crown of the plant remain for feeding on the farm, to which can be added the pulp, a residue of processing. Sugar beet has been preferred as fodder to fodder beet, perhaps partly because of its alternative uses. It is also valuable for the farms in the rotation as an intertilled (*propashnoy*) crop alternative to other roots.

Much of the advantage of cultivating sugar beet as against other crops is lost if the processing industry is weak. Not only must there be a satisfactory and continuous balance between the number and capacity of factories on the one hand and the area under the crop in the supplying region on the other, but transport services must be oriented to the industry. These desirable relationships have not yet been attained in the U.S.S.R. There is overcapacity of factories in some areas, undercapacity in others, and an average length of haul of beets to the factory of 206 kilometres in 1958 (a year of high procurements) and 155 in 1962, when procurements were below average.[1] Excessive length of the processing season is also characteristic of the Soviet sugar-beet industry.[2]

VEGETABLE OILS

Vegetable oils are derived from several crops in the Soviet Union. As secondary products, oils are obtained from cotton, flax and hemp, but more important are the specialist crops—which, nevertheless, also yield fodder—the chief of these being sunflower.

The sunflower demands moderate warmth—about 2,200 day-degrees—but is resistant to drought and hence can be widely cultivated in the steppes. In parts of the western steppes it is among the most important crops, occupying 10–15 per cent. of the sown area.[3] It covers significant proportions of the land from the Ukraine and central black-earth region to the north Caucasus, Transcaucasia and western Siberia. The area sown with sunflower has increased from under one million hectares before the Revolution to about 5 million hectares annually since 1965, of which rather over half is in the R.S.F.S.R. The all-Union average of 13·7 centners per hectare in 1968 was nearly double that of 1940, giving a total harvest of 6·7 million tons compared with 2·6 million in 1940 and 750,000 tons in 1913. This is nearly 90 per cent. of the total 1968 yield of oilseed crops. Furthermore, plant-breeding has evolved much more productive strains so that the oil yield at the factories is about 43 per cent., compared with 25·4 per cent. in 1940.[4] Continued selection of plants has produced still higher yielding strains and varieties which ripen in 12–14 days less than previous high-yielding varieties.

The soya bean is a possible competitor with the sunflower in the warm, dry steppes but is not widely cultivated because, with limited drought resistance, it needs moist soils, especially in summer. Yield is reduced by low night temperatures, and furthermore it is frost-tender. Nevertheless, it is a valuable

[1] Hultquist (1967), 139–140.
[2] Hultquist (1967), 141–142.
[3] Nikitin and others (1966), 170.
[4] Suslov (1967), 118.

source of protein as well as of vegetable oil, and will probably spread further. Total soya area was expanded from 290,000 hectares in 1940 to 848,000 hectares in 1965, of which 844,000 were in the R.S.F.S.R.—nearly all in the Far East, particularly in the Amur, Khanka-Ussuri and Zeya-Bureya lowlands. Three-quarters of the area was on state farms. Most of the non-R.S.F.S.R., sowings are in Georgia and Moldavia.

Other oilseeds are grown in small quantities. Mustard is grown in the steppe areas, notably Volgograd, Rostov and Saratov oblasts and in north-west Kazakhstan, in all about 290,000 hectares. Resistant to drought, it can be cultivated in semi-desert land. About 170,000 hectares of castor oil plants are grown.

Winter rape, capable of yielding 8–10 centners of oil per sown hectare, offers possibilities in areas where other oilseeds are not practicable, such as the cool lands of Belorussia and the Baltic republics. Linseed, hemp and groundnuts are other oil-bearing plants grown in various areas, the last probably only experimentally. The residue of linseed is familiar in other countries as a cattle-cake, and if the oil is not extracted the seed is nutritious. Indeed, most oilseed plants have a useful place in livestock husbandry. Tung, however, grown on a small scale in western Georgia for its use in paints and varnishes, has a poisonous residue. Also requiring warmth, but less moisture, sesame is grown in other parts of Transcaucasia, and in Central Asia and south Kazakhstan.

FLAX

Flax is important in the U.S.S.R. Although linseed is produced from the bushy varieties grown in the south from the Ukraine to Central Asia, flax for fibre is much more important. Both fibre and seed are produced in the forest zone. Growth begins early in the season and can be accommodated readily within the short summer as far north as 60 degrees. Flax is tolerant of a wide range of soils, but not of very light, sandy textures. Weeding is important, and because of the risk of disease, repetitive sowings are to be avoided, so that not more than 10–15 per cent. of the sown area is used for flax, although it is very profitable. Fine quality fibre is produced in the very northerly areas, but the podzols and gley soils there need considerable fertilizing, so raising costs. The soils of Pskov and Novgorod oblasts also demand special treatment, but the climate of the north-west ensures minimum risk of dry conditions at the beginning of the growing period, which is critical for fibre production. It is a nutrient-exhausting crop but fits in well with potatoes and other roots, and dairying and other livestock enterprises of the area.

Yields of flax have averaged 3·3 centners per hectare in recent years compared with 1·3 in 1950. The Soviet statistics accept the yield of 3·3 for 1913, which makes present performance look poor, but almost certainly the performance of 1913 has been improved upon. A discrepancy could be explained by statistical errors or fibre improvement concealed in gross yields. Between 1965 and 1969 the U.S.S.R. sown area diminished from 1·48 to 1·31 million hectares, compared with 2·1 million in 1940 and 1·25 million in 1913. The U.S.S.R. is by far the world's largest producer of flax and it is an important export, but the area will probably be progressively reduced as demand for linen products falls in competition with man-made fibres. Labour demands are seasonally heavy in flax cultivation, and mechanization has been only moderately successful in reducing these demands.

HEMP

Hemp cultivation overlaps that of flax, but even the northern varieties need 100–110 days for growth while southern hemp needs 120–160 days. The hemps are, therefore, mainly crops of the wooded steppe and are prominent on the interfluves of the central black-earth region—Orel oblast and neighbouring parts of Kursk, Tula and other oblasts. The belt of cultivation extends westward to Belorussia, the western Ukraine and Moldavia, with increasing importance of southern varieties and with lower density—generally below one per cent. of the sown area. Eastward, beyond a gap around the upper Don, another area of concentration extends from the east of Tambov oblast through Penza oblast, thinning out as the Volga bend is approached. More isolated areas of hemp cultivation occur further east. Southern hemp is grown in the north Caucasus area, Kirgizia and southern Kazakhstan as well as in the Ukraine and central black-earth region.

Soviet plant breeders have assiduously attacked the problem of producing seed and fibre from the same plant, both products being objects of hemp production but normally from different plants. Some success is being achieved in this as well as in developing more productive and faster growing varieties.

The total area sown to hemp has been much reduced, from 680,000 hectares in 1913 and 600,000 hectares in 1940, to 230,000 hectares in 1968. The R.S.F.S.R. total was 514,000 hectares in 1913, 357,000 in 1940 and 138,000 in 1968, of which all but 10,000 hectares was on state farms. Improved yields, however, have raised production from 69,700 tons of fibre in 1940 to over 90,000 tons in 1966. Average per-hectare yield rose to 4 centners in 1966 compared with 2·4 in 1940, but many hemp-growing collectives achieve over 9 centners per hectare.[1]

KENAF & JUTE

Kenaf and jute are two relatively recent introductions, both requiring warm climates for their development. Central Asia and Transcaucasia offer possibilities for their development and both are grown on a few farms. Irrigation and improvement of mechanization of handling are essential prerequisites for their economic development, but, the early problems having been at least partly overcome by research, the Soviets claim a world record for yield of kenaf fibre of 190 centners per hectare on an Uzbek farm.[2]

COTTON

The most important fibre crop in the U.S.S.R. is cotton, which is, of course, confined to the warmer areas, (Figure 11), and now almost completely to irrigated lands. Growth begins only when temperatures are about 15 degrees C, and the optimum is between 25 degrees and 30 degrees, with much bright sunshine. Growth slows at temperatures below 17–20 degrees, and even light frost is fatal. The crop is drought-tolerant but regular water supply is needed for a good harvest, so irrigation is necessary in the parts of the Soviet Union otherwise most suited to the crop. Two-thirds of the sown

[1] Senchenko (1967), 119.
[2] Senchenko (1967), 120.

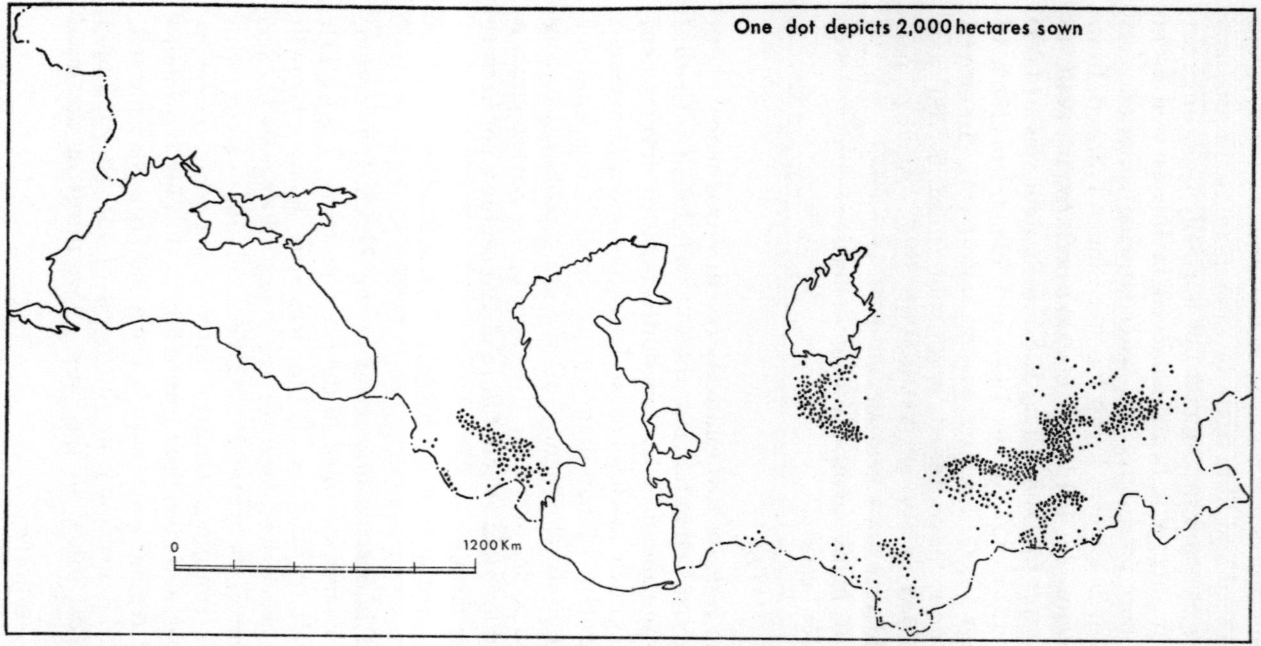

11. Cotton: distribution of cultivation, 1964

The areas devoted to cotton have also been substantially increased since the war but in this case the increase has been confined to the areas most suited to the crop—Central Asia and Transcaucasia—and attempts to grow it in the Ukraine have been abandoned

Source: *Atlas razvitiya khozyaystva i kul'tury SSSR*, 66

area is in Uzbekistan, while the Central Asian republics and Kazakhstan together contain 90 per cent. of the area, the other 10 per cent. being in Transcaucasia.[1]

During the inter-war years cotton growing was extended to non-irrigated lands of the southern Ukraine, north Caucasus and Volga delta. In 1938 these areas accounted for one-quarter of Soviet cotton lands, but the marginality of the climate for the crop was reflected in poor yields and, in some years, almost total failure of the crop.[2] Attempts to grow cotton in these unsuitable areas were renewed after the war but dropped in favour of concentration on increasing production on the irrigated lands of Central Asia and Azerbaydzhan. Cotton was shown as occupying 124,000 hectares in the non-irrigated areas in 1955, compared with 526,000 in 1940, while in the irrigated areas it had expanded from 1·5 million hectares in 1940 to 2·1 million in 1955.[3] The area in Central Asia and Transcaucasia was increased slowly and reached 2·54 million hectares in 1969.[4] Average yields rose only marginally from 13 centners per hectare in 1909–13 to 14 centners in 1948–52, then for 1953, with less non-irrigated cotton and more fertilizers available, a yield of 20·5 centners was reported. Further improvements brought the figure to 24·5 in 1967, giving an average for 1964–68 of 23·5 centners per hectare. The 1969 yield, however, fell to 22·5 centners per hectare.[5] Total production has almost touched 6 million tons per annum, eight times as much as in 1913, on four times as much land. The U.S.S.R., is still, however, far from self-sufficient in cotton and further expansion of cotton growing is planned. This expansion will be mainly in Central Asia, yields in Transcaucasia being very much lower. Azerbaydzhan, however, retains special significance for long-staple cotton.

TEA & TOBACCO

Also crops of the warm areas, but demanding less heat and sunshine than cotton, tea and tobacco are grown mainly in the Caucasus and other southern regions. Western Georgia,

[1] See Chapter 10, region 20.
[2] Volin (1951), 138–142.
[3] *N.kh. SSSR 1960*, 401.
[4] *N.kh. SSSR 1969*, 318.
[5] *N.kh. SSSR 1969*, 332.

adjacent areas of Krasnodarsk kray and the Lenkoran region are the main areas of tea production. A little over 70,000 hectares grow tea, producing about 230,000 tons annually.

Whereas tea can withstand light frosts, tobacco is very frost-tender, and is hence found on lower and milder lands, where rainfall is high and soils are leached. The Crimea and other Ukrainian areas, the western Caucasus, Transcarpathia and Moldavia are the main tobacco areas, but it is now fairly widespread, extending to Central Asia. Makhorka, the strong traditional pipe tobacco, is grown widely in the black-earth and Volga areas and western and central Siberia.

State purchases of makhorka fell from 168,000 tons in 1940 to 39,000 tons in 1969, whereas tobacco purchases rose from 73,000 tons in 1940 to 215,000 tons in 1968 and 195,000 tons in 1969.[1] Tobacco is, of course, an intensive crop and one Moldavian kolkhoz has earned over 7,000 rubles per hectare with yields of 32 centners.[2]

POTATOES

Potatoes are important in the U.S.S.R., being used for industrial purposes (starch and alcohol) and animal fodder, as well as human consumption. The modest requirement of 1,000–1,300 day-degrees C. of accumulated temperature in the growing period (only some 70 days for fast-growing varieties) enables them to be grown further north than the fastest growing grains. They are tolerant of a wide range of humidity, though for the best growth considerable soil moisture is necessary late in the growing season. Waterlogged soils, however, encourage diseases of the tubers and hinder harvesting. Tolerance of acid conditions facilitates potato cultivation in the forest region.

In 1913, 4·2 million hectares were under potatoes. By 1940 the area was 7·7 million and after the war it rose to over 9 million hectares. There has since been some contraction—in the period 1965–69 from 8·6 down to 8·1 million hectares.

Yields vary greatly with weather and other local conditions. The highest U.S.S.R. average claimed was 123 centners per hectare in 1968.[3] and the three-year average 1966–68

[1] *N.kh. SSSR 1969*, 299.
[2] Leonov (1967), 124.
[3] *N.kh. SSSR 1969*, 340.

was 114 centners. The high yields of 1967 and 1968 enabled record production figures to be claimed despite the drop in area. At 102 million tons the 1968 figure was more than three times the 1913 figure.

Potato production is led by the cool and humid north-western areas. For some years past, Estonia has consistently reported the highest average yields, amounting in 1968 to 190 centners per hectare, when the U.S.S.R. average was 123. Above-average yields characterize the other Baltic republics, Belorussia and the north-west of the R.S.F.S.R., which, with the northern Ukraine, are the main potato-producing areas. The lowest yields are in the southern areas from Moldavia to Central Asia, where high temperatures and lower humidity adversely affect the crop, and virus diseases are common. Potatoes have been a staple food in European Russia for centuries and, though, with higher standards of living, they show some decline as a food, they remain important in the north and west. Understandably, farms try to market potatoes for human consumption to obtain the highest prices. In this trade there is now some specialization, some farms near cities having 1,500 hectares in potatoes.[1] Potatoes also are used as fodder raw, cooked and in silage, and are fed to cattle, pigs, sheep and poultry. They can acceptably form up to 45 per cent. of succulent food for cattle and over 70 per cent. of pig rations. The use of potatoes in alcohol production has greatly decreased as compared with pre-war years,[2] but industrial outlets are still important in the north of the wooded steppe and the southern forest zone.[3]

VEGETABLES

Vegetable production is also still largely by individual collective farmers on their personal plots, but less so than formerly. Large state farms specializing in vegetable growing have been created around Moscow, Leningrad, Kiev, and other large cities and have considerably improved the supply position. Cultivation under glass has become more important. Warmth-demanding vegetables, however, come to the cities

[1] Dorozhkin (1967).
[2] Volin (1951), 128.
[3] Nikitin and others (1966), 174.

from further afield, specialization in these having been developed in the southern regions. These measures have enabled the period when fresh vegetables are available fairly generally in the big northern cities to be increased to six months or so.

The total area sown with vegetables, excluding potatoes and melons, has fluctuated little in recent years from 1·4 million hectares, just over twice the reported area of 1913 but similar to that of 1940. Yields have increased to an average of 132 centners per hectare (1966–68), from 73 centners per hectare in 1950–52 but these figures are of little value, covering as they do such a great variety of products and purporting to include the yields of all the many personal plots. State purchases mean a little more, and at about 9 million tons (1967–68) per annum are currently over four times the 1950 figure (2 million tons)[1].

Melons are important in the southern areas. Warmth, bright sunshine and ample water are necessary for their cultivation, and so, with irrigation, they are grown in Central Asia for export to other regions. Similarly, water melons are sent to other areas from the lower Volga (especially Astrakhan and Volgograd oblasts), the north Caucasus and southern Ukraine. Fodder melons and fodder pumpkins are a valuable part of the animal feed of the steppe zone. These are referred to later.

In the whole U.S.S.R., about 600,000 hectares are under melons, excluding fodder melons, an area which has changed little since the war, though twice as great as in 1913.

FRUIT & VINES

Fruit growing is widespread on collective farms, both as a collective enterprise and on personal plots. Climatic considerations, however, restrict the kinds that may be grown in the northern areas to the hardier temperate fruits, with apples most common. Pears, plums and soft berries are also popular. Concentrations are found near Moscow, and other large cities. Cherries and apricots are among the more varied

[1] When, in 1950, state purchases were reported as 2·04 million tons, total production was estimated as 9·3 million tons. In 1968 total production was estimated as 19·0 million tons, and state purchases were 9·0 million tons, i.e. the state purchases appear to have more than quadrupled while the output has doubled.

fruits grown south of Moscow. In the western Caucasus and on the Black Sea coast, citrus fruits have been grown since the twelfth century. The Caucasian republics produce a wide variety of fruits, including olives, persimmon, peaches, apricots and almonds and other nuts. Central Asia is noted for cherries, apricots and other fruits, while the name of Alma Ata itself advertises the fine, large apples grown there.

Grapes, both table and wine varieties, are grown in Georgia, the Crimea and elsewhere in the southern Ukraine, Moldavia, the north Caucasus and Central Asia.[1] Relatively dry and sunny areas with well-drained stony and sandy soils are, as elsewhere, preferred for viticulture.

From only about 650,000 hectares in 1913, the area in orchards of all kinds was increased to 1·6 million hectares in 1941. Nearly one-tenth of this area was destroyed during the war, but recovery and further expansion took the area to 3 million hectares in 1956 and to 3·6 million in 1965, since when there has been little change. This area is divided about equally between state farms, collective lands and personal plots. The R.S.F.S.R. and Ukraine each contained about 1·3 million hectares of orchards, but the proportion in state farms in the R.S.F.S.R., 45 per cent., was more than twice as high as in the Ukraine.[2]

CITRUS FRUITS

Various citrus fruits are grown in the U.S.S.R., but the vulnerability of these fruits to frost limits them to the warmer areas. The country has been divided into three zones of suitability for citrus trees. The first zone, where the mean annual minimum temperature does not fall below −3 degrees C., is considered suitable for growing lemons in the open; in the second, mean minimum −3 degrees to −4 degrees C., lemons should be restricted to sloping fields and mandarins and oranges should be sheltered; and in the third zone (−4 to −6 degrees C.) lemons may still be grown on sloping ground but for mandarins and oranges dense stands as well as shelters are required. With mean minimum temperatures

[1] Mellor (1964), includes a brief but informative account of the Soviet wine industry (pp. 209–210).

[2] *N.kh. SSSR 1968*, 374–375.

below −6 degrees, citrus fruits should not be planted.[1] In practice, citrus trees are concentrated on the Caucasian coast of the Black Sea, where about 10,000 hectares are planted.

FODDER CROPS & GRASS

The most striking and significant change in the use of the arable area in the U.S.S.R. during the past half-century, particularly during the last two decades, has been the remarkable upsurge in the growing of fodder crops. In 1913 only 3·3 million hectares were recorded as carrying crops for fodder, mostly perennial grasses. By 1950 the corresponding figure was 20·7 million hectares, rising to 50·2 million hectares in 1958 and 63·1 million in 1963. More recently, the fodder crop area has fluctuated, the area in 1965 being 55·2 million, and in 1968, 60·7 million hectares. The figure of almost 30 per cent. of the sown area so used has been typical of recent years.

Realization that livestock needed good fodder to carry them through the long winter led to small-scale promotion of fodder crops before the Revolution, including the establishment of research stations, principally concerned with this aspect of husbandry, in 1910. After the Revolution a State Meadow Institute was formed, this becoming in 1930 the All-Union Scientific Research Institute of Fodder Crops. At this time, the 500 million hectares of grasslands termed 'natural' and virtually so, except for grazing pressure, provided the main base for the livestock industry. Even in 1956, natural pastures still provided 75–80 per cent. of all hay.[2] The yield of the unimproved lands was, however, low and farms were pressed to plough-up old and low-yielding pastures or to improve them by drainage, fertilizing and other methods.

Introducing grasses into rotations on arable land was promoted under the *travopol'ye* system, which became standard rotational practice in 1937, enthusiastically backed by Stalin.[3] Khrushchev became disillusioned with it and eventually denounced it vigorously in the decree of March, 1954, which

[1] Brezhnev and Minkevich (1958) 1961, 71.
[2] Badiryan (1956) 1960, 96.
[3] As described in Chapter 3.

urged the growing of more feed grains and succulent crops and the development of virgin and long-fallow lands.[1]

It is easy to criticize the Vilyams system, and criticism for undue concentration on one element in rotations, grass, is justifiable. Nevertheless, it must be remembered that, by whatever method is practicable, fertility must be restored after harvesting exhausting crops such as wheat. Fallowing is probably necessary in many dry regions to conserve moisture for the grain crops as well as to rest the land, but wherever productive use of the land is feasible, this must obviously be preferred to bare fallow. Grass leys can support livestock, which, through their manuring, promote the nitrogen cycle. Furthermore, clover or lucerne included in grass mixtures effect nitrogen fixation, so that, even when the crop is taken off as hay, land improvement results.

Clover offers these possibilities for the more northerly and cooler parts of the Soviet Union, while lucerne is more advantageous in the drier areas, where its long roots can maximize use of the soil water resources. The term *mnogoletnye travy*, usually translated as 'perennial grasses', includes clover and lucerne. Lucerne is tolerant of saline soil conditions and so can also be used on the semi-desert soils. Seed is, in fact, raised in Central Asia, where lucerne is grown on irrigated land in rotation with cotton. Timothy grass is much used in mixtures for heavy land in the cooler areas, particularly for hay.

Of the Soviet Union, as the cult of travopolye was ending, a geographer commented:

> One can only agree that the ratio of perennial to annual grass is far too low. The recent trend in acreages devoted to these crops seems to indicate that Soviet planners realize this fact. The 1962 cutbacks reduce annual grasses to a much greater extent than perennial grasses and, in fact, restore the latter to the dominant position which they occupied among hay crops prior to 1956.[2]

The 1962 figures showed, in fact, 15·6 million hectares under perennial grasses and 11·7 million under annual grasses,

[1] *Pravda*, 6.3.54.
[2] Anderson (1963), 250.

respectively 4·2 million and 5·0 million hectares below the 1961 figures.[1] The main increases over 1961 figures were in maize and legumes. In fact, however, with decline in the maize area and other alterations, sowings of annual grasses increased again and the 1966–68 average was 17·5 million hectares. The reduction in area of perennials continued until 1964, when, at 12·6 million hectares, they occupied only 65 per cent. of the 1961 area. Renewed attention to perennials was ordered in 1966 and their area increased to 19 million hectares in 1967.

Excluding the grains, the proportions of fodder crops to sown area in 1967 were as shown in Table 14, with the 1960 figures for comparison.

TABLE 14

FODDER CROPS IN RELATION TO SOWN AREA

	1960		*1967*	
	Million hectares	*Per cent.*	*Million hectares*	*Per cent.*
Total sown area	203·0	100·0	207·0	100·0
of which, fodder crops	63·1	31·2	60·7	29·1
of which, annual grass	19·3	30·8	18·0	30·0
and, perennial grass	16·8	26·6	19·0	31·4
and, maize silage and fodder maize	23·1	36·6	19·0	31·4

Source: N.kh. SSSR 1968, 334.

The regional changes necessary to achieve a high proportion of the land under fodder crops are shown in Figures 12 and 13. During the period to 1958, the outstanding increases in the sowing of fodder crops were in Kazakhstan and south Siberia, mainly in connection with the virgin land programme, although the principal object of that development was to increase wheat production. Similar increases in terms of ratio, though not of volume, were attained in Belorussia, where the increase in area under fodder crops more than offset the decline in the grain area. The central zone from the Urals to the north Caucasus showed intermediate rates

[1] The figures given here are the final official statistics. At the time of his comment Anderson had only figures from *Pravda*, 21.7.62, but the comment requires no alteration.

of increase. The greatest regional increase of all in volume, however, occurred in the Ukraine, although on the map this is not obvious because of the lower rate of development from the already relatively large base of fodder crops. In the R.S.F.S.R., major proportionate increases occurred not only in the regions already mentioned but also in east Siberia,

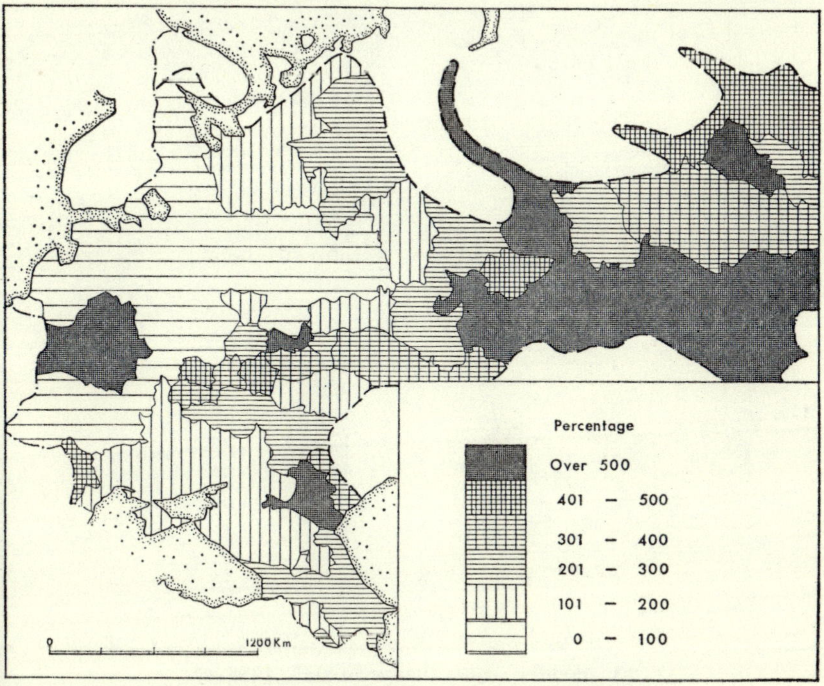

12. Fodder crops: increase in area, 1940–58

the Far East and the north, but these zones had only small areas involved. Although the map compares the 1958 position with 1940, the increases took place mainly in the 'fifties, particularly after Khrushchev's policies became effective.

In the following period, 1958-65, consolidation of the position was the theme and only modest increases in fodder crops were recorded in the virgin land areas and there were decreases in some of the areas of European Russia that had previously recorded increases. In most central and southern areas, however, there were moderate increases leading to a more balanced distribution of fodder growing required for the advance of livestock husbandry.

The aim of providing animals with an uninterrupted supply of green and succulent fodder throughout a grazing season extended by suitable crops in spring and autumn has become

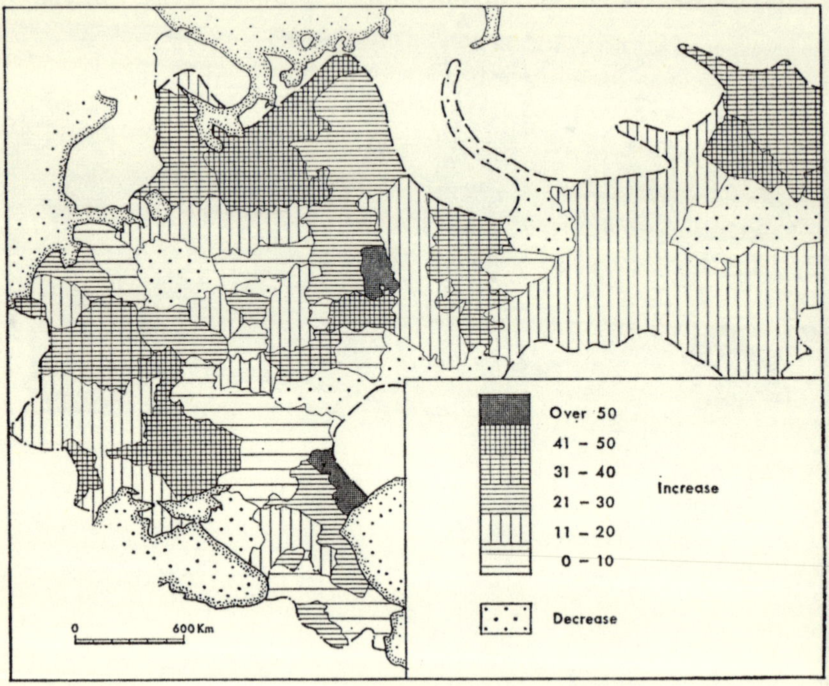

13. Fodder crops: change in area, 1958–65

known as the 'green conveyer system'. Appropriate crops are selected according to regional characteristics.

There is, of course, great variation in the periods for which livestock have to be maintained by stall feeding in the different climatic and vegetational zones. Examples are given in Table 15.

The percentage of land under sown fodder crops is at its highest in the north-west of the U.S.S.R., being regularly between 40 and 50 per cent. of the sown area in the three Baltic republics and the North-west region. This was true, for example of 1967, when the Central, North Caucasus,

East Siberia regions, and the Ukrainian, Kirgizian and Armenian republics also had over 30 per cent. in fodder crops.[1]

Considering the grassland area expressed as a percentage of total sown area, the Baltic republics and North-west region recorded over 30 per cent., the Central region, Kirgizia and Armenia over 20 per cent., but East Siberia and the Ukraine, as might be expected, did not have a distinctively high grassland area.[2]

TABLE 15

PERIODS OF STALL AND PASTURE FEEDING IN SELECTED ZONES

Zone	*No. of days per year*	
	Stall feeding	*Grazing*
Forest and forest-steppe	185–235	180–130
Steppe	about 165	about 200
Semi-desert	115–145	250–220
'Desert'	—	all year

Based on figures quoted by Ignatov (1956), 253.

There is little difference between the figures for the whole U.S.S.R. and for the R.S.F.S.R., containing, as it also does, areas so differently suited to grass. In 1965, the U.S.S.R. fodder area was 26·4 per cent. of the sown area, that of the R.S.F.S.R., 27·1 per cent., while the sown grassland areas were 14·3 and 14·6 per cent. respectively.

The attractions of grass are illustrated by costings which have shown that careful husbandry and control of stock on enclosed pastures enables Estonian farms to obtain 3,000–4,000 feed units at a cost only 20 to 30 per cent. of that of other fodder crops. Such achievements are only possible in moist areas. In the dry regions of the Volga area, however, mixtures of grasses and alfafa (for example *Bromus inermis*, *Festuca pratensis* and alfafa, *Medicago sativa*) in water meadows in Saratov oblast yielded 111 centners per hectare, and in another experiment in the south-eastern Transvolga area *Bromus inermis* and alfafa yielded 145 centners per hectare in 1952.[3]

[1] *N.kh. SSSR 1967*, 350–351, 354, 366.
[2] *N.kh. SSSR 1967*, 354, 367.
[3] Brezhnev and Minkevitch (1958) 1961, 99.

HAY

Of the 40·1 million hectares of natural hayland in the U.S.S.R. in 1968,[1] 27·5 million were in the R.S.F.S.R. Only in the North-west region did the area closely approach the sown area (2·6 million compared with 2·8 million hectares sown). In the Far East it was 1·8 million compared with 2·6 million sown. In the North Caucasus the ratio of natural hay to sown area was less than 1:12, in the Central Chernozem region only 1:19.

The 1965–68 U.S.S.R. averages for hay from annual grasses was 13·1, while perennial hay yielded an average of 17·2 centners. Both these yields were better than all but exceptional years before 1966. They were also, as they should be, much better than the yields from natural pasture averaging only 5·5 centners per hectare from 1965 to 1968.

According to the statistics, the yields of natural hay have shown a long-term decline, and until recently, post-war hay yields from sown grasses also compared poorly with earlier periods.

Jasny commented:

> By immense negligence hay yields per hectare were permitted to deteriorate greatly in the U.S.S.R. Wild or natural hay yielded only 7·1 quintals (centners) per hectare in 1959 as against 13·4 quintals in 1928. Perennial and annual hay returned 16·8 and 11·8 quintals respectively in 1959 as against 29·1 quintals for all rotation hay in 1928.[2]

He explained this by delays in cutting the hay, cutting too high, ploughing-up the best meadows and shifts of hay production to the lower-yielding eastern areas, and he concluded that Soviet yields of hay 'may be the lowest in the world.'[3]

Commenting on these observations, Anderson[4] agreed that late cutting could be one reason for the low yields (a fault, incidentally, common in many countries). He thought high

[1] Excluding 5·7 million hectares of haylands in *goszemzapas* and forest organizations and 1·1 million hectares in the hands of 'other users'.
[2] Jasny (1963), 220.
[3] Jasny (1963), 221.
[4] Anderson (1963), 252–253.

cutting no longer very important, but considered very relevant the sharp drop in manure available. This initially arose through the drastic decline of livestock following collectivization,[1] and more recently, to increased stall feeding, whereby more manure was diverted to cultivated crops or wasted.[2] He also noted the decline of floodplain meadows from 16·6 million hectares to 10·4 million hectares by 1955,[3] though the total area of hay lands had remained nearly constant. Yet another possible reason cited was the tendency to prefer to spend time on more valuable crops both in the collective fields and on personal plots. Inadequate mechanization of hay-making was also relevant—in 1959, 44 per cent. of hay cutters and 70 per cent. of hay rakes being horse-drawn, whereas no other horse-drawn machines were enumerated.[4]

Hay yields are in fact particularly difficult to compare both in space and in time, because of varying species in the herbage and their different weights and values as fodder, besides the normal lack of need to weigh the crop, in contrast to cash crops. Estimating the crop is easier when hay is baled by machine, but this was not normal in the U.S.S.R. in the period under discussion. Furthermore, the amount of grass that can be taken for hay is conditioned by the grazing pressure. The labour expended in gathering poor quality 'natural' hay may be uneconomic if the grass can be grazed and the winter feed be provided by crops, silage and concentrates.

We should also note that hay yields in most of the U.S.S.R. can hardly be expected to compare with those of western European countries or many parts of North America, which enjoy cool, moist climates much more conducive to the growth of grass suitable for hay, even though the same climates may hinder its harvesting. These points do not apply to sown meadows or lucerne stands, which should be related to the potential of the local soils and climate, but one still needs to know more about the grassland management system in use before being too condemnatory of recent trends.

[1] This would not be relevant to the post-1950 period, as livestock numbers had recovered to above pre-collectivization levels.

[2] This should be less necessary with improving supplies of mineral fertilizers and handling-machinery.

[3] Anderson quotes authorities but it is not certain that these figures are on the same basis.

[4] *S.kh. SSSR* (1960), 415.

PASTURE LANDS

Transferring attention from land use aimed mainly at production for direct human consumption to purely livestock-rearing uses is complete when we consider natural pasture. Unlike the sown fodder lands, the natural pastures offer no alternative uses unless they can be changed completely in character by ploughing and reseeding.

Of the whole Soviet territory, 14·7 per cent. (326 million hectares) was recorded as in pasture in 1968, excluding the very low productivity reindeer pastures of the far north. Of this area, 36·8 million hectares were held by state and forest organizations and not in long-term use by collectives or state farms, and 17·2 million were classed as 'other lands'. Of the remaining 272 million hectares, 141 million were in Kazakhstan, 61 million in the Central Asian republics (mainly in Turkmenistan and Uzbekistan) and 58 million in the R.S.F.S.R. In the economic regions of the R.S.F.S.R., the highest proportion of pasture was found in East Siberia (45 per cent of the total agricultural area). Next were the North Caucasus and Volga regions with 32 and 30 per cent. respectively. The Far East, Ural and West Siberia regions were close to the R.S.F.S.R. average of 25 per cent., the North-west region averaged 21 per cent. and the lowest proportions were in the Central, Volga-Vyatka and Central Chernozem regions (15, 15, and 12 per cent. respectively).

It is also noteworthy that of the 272 million hectares of natural pasture in the U.S.S.R., 183 million or about two-thirds, are in state farms. This high proportion, however, is almost wholly explained by the inclusion of 122 million hectares of Kazakh pastures in state farms, compared with only 19 million in collectives. Much of the pastoral land of Kazakhstan is particularly poor, virtually semi-desert, on which there is still an element of nomadism in the livestock herding, but the amount in state farms has been increasing, both relatively and absolutely, in recent years.

Exclusion of Kazakhstan from the figures at once more than halves the total pasture area to 131 million hectares, of which 60 million (46 per cent.) were in state farms in 1968. For the R.S.F.S.R., with a total pastoral area of 58 million hectares, the state-farm portion was about 55 per cent.—over 60 per cent. in the Far East, West Siberia and Ural regions.

The statistics show all natural pasture land as held by either state farms or collectives except for the state forest and 'other' users, i.e., personal grazing was not distinguished, being carried out on lands belonging to the collectives, if not on the actual personal plots.

TABLE 16

AVERAGE TOTAL AREA REQUIRED TO SUPPORT ONE LIVESTOCK UNIT

	Hectares
Central Chernozem, Ukraine, Moldavia, Georgia, Armenia, Azerbaydzhan	1
Belorussia, Baltic republics, non-chernozem belt	2–4
Siberia	4–6
Kazakhstan and Turkmenistan	12+

Based on figures quoted by Zaltsman (1956) 1960, 230.

If the areas of barley, oats and legumes are added to those classified as fodder crops, grass, hay and pasture, the total area devoted to livestock feeding is approximately 404 million hectares, nearly 75 per cent. of the total agricultural area. Of this livestock feeding area, 272 million hectares, 68 per cent., are natural pasture of low value. In Turkmenistan, 99 per cent. and in Kazakhstan, 90 per cent. of land available for livestock feeding is natural pasture. These figures stress the desirability of upgrading natural pasture lands by irrigation, topdressing, reseeding, drainage or other appropriate methods as discussed in Chapter 6.

Flood meadows constitute only about 10–12 per cent. of the 'natural' grasslands but have yielded 20–25 per cent. of hay harvested.[1] Such meadows may approach the chernozem lands in potential fertility but the average feed produced per hectare is only one-third or one-quarter that from the average chernozem.[2]

The high proportion of poor pasture is apparent in the large average total area of land required to support one livestock

[1] Badiryan (1956) 1960, 97.
[2] Badiryan (1956) 1960, 98.

unit (one cow or equivalent other stock) on all but the most fertile areas as illustrated in Table 16.

Though productivity of pasture in the U.S.S.R. is low, its large total area ensures great importance in the total feed supply for livestock. This is shown in Table 17 which also

TABLE 17

EXPENDITURE OF FODDER ON CATTLE AND POULTRY
(million tons)

	All types of holdings		*In which, collectives and state farms*	
	1964	*1967*	*1964*	*1967*
Concentrates	48·2	75·1	33·6	53·7
Succulent fodder	361·2	412·0	280·1	308·5
in which, silage	144·2	147·1	137·3	139·3
Roughage	142·3	166·9	94·8	112·5
in which, hay	69·8	78·4	41·9	48·4
Pasture feed	347·2	416·8	238·6	274·3
All fodder, in fodder units	239·8	295·0	167·2	202·9

Sources: N.kh. SSSR 1964, 377; *1968,* 111.

illustrates the difference in the importance of pastoral and non-pastoral fodders in 1964, following bad harvests, and in 1967, after a much better harvest. The differences in feeding of the collective and state herds, compared with the total, which includes private livestock feeding, are also noteworthy, if to be expected. The former utilize proportionately much more silage, less hay, rather less concentrates, and pasture makes about the same contribution to the total.

CHAPTER 9

The Livestock Industry

In Russian agriculture livestock have traditionally played a role subsidiary to that of cropping. In some other territories of the Soviet Union, however, livestock have been of greater relative importance through the ages because of the mountainous or arid conditions as, for example, in Central Asia, Kazakhstan, southern Siberia and the Caucasus.

The serf system, the smallness of the holdings granted to the serfs when they were freed, their poverty and the three-field system and communal management all militated against the evolution of good stock raising. The First World War, the collectivization drive and the Second World War successively resulted in vast losses of stock, while in the intervening years shortages of fodder and skilled management delayed the recovery from each preceding catastrophe.

As long as the country remained in danger of famine the production of food by the indirect and relatively slow and costly processes of animal husbandry had to be placed second to cropping. Until mechanization had advanced enough for draught animals to be no longer needed in great numbers, improvement of breeds for food production was hindered, and until payment for work with livestock was improved the members of collective farms were interested only in their small numbers of stock owned privately.[1] As Soviet agriculture has overcome these obstacles, however, it has gradually become possible to attack the fundamental problems facing the stockbreeders.

In the recent drive to increase agricultural productivity a substantial effort has been directed at the livestock branches and conspicuous gains have been recorded. The development of animals suitable for the varying conditions in the vast

[1] Volin (1951), Chapter 8.

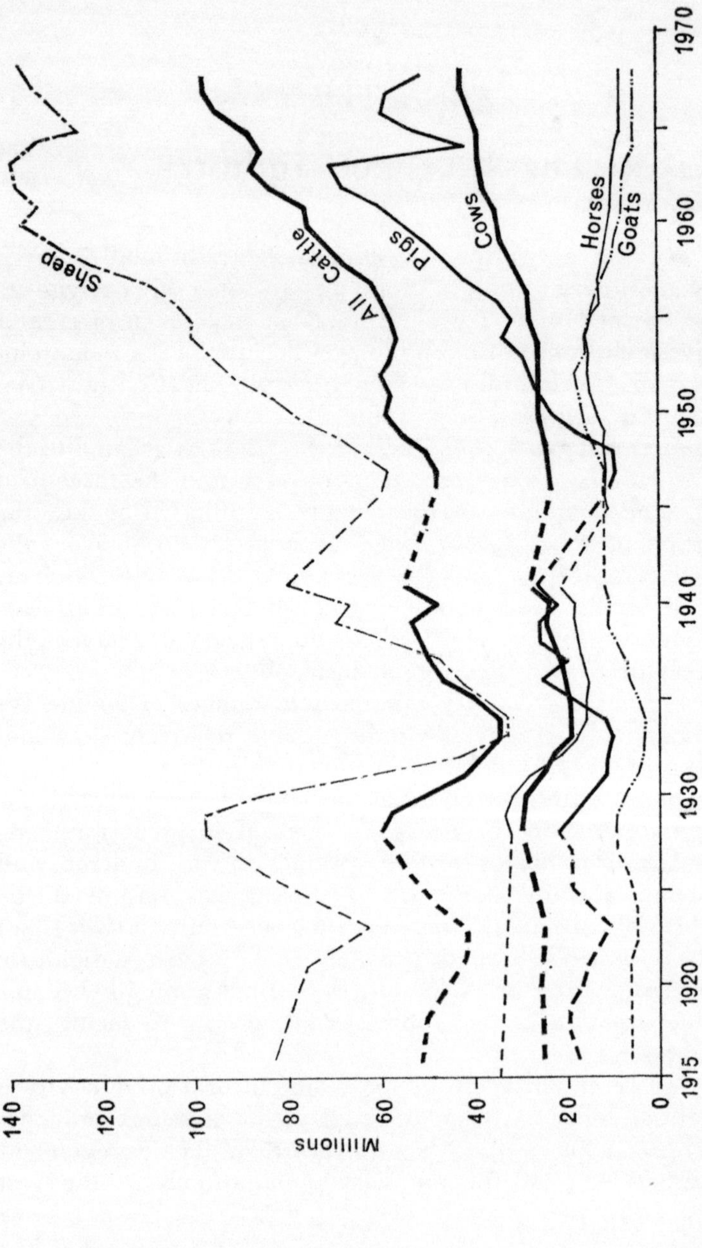

14. Changes in the numbers of the main classes of livestock

area of the U.S.S.R. has been emphasized, and much use has been made of imported pure-bred stock for crossing with native breeds to develop the desired qualities. Most of this work has been done in scientific establishments and on selected state farms, but collective farms have also made substantial contributions.[1] To establish a breed takes up to 25 years and the pre-war Soviet work on breeding beef cattle involved the importing of over 3,000 pure-bred Hereford, Shorthorn and Aberdeen-Angus cattle from Great Britain and Uruguay.[2]

The main effort in breeding livestock in the Soviet Union has been organized through state breeding grounds and breeding sovkhozes. The staff of the state breeding grounds supervise the kolkhoz farms in surrounding areas. They attend to breeding, feeding, grading, registration in herdbooks, exhibitions and other aspects of breeding work.[3] Breeding sovkhozes produce mainly pure-bred cattle of high quality.

Herdbooks are published by the supervising body, the State Department of Inspection for Breed Improvement and Herdbooks. Standards are laid down specifying minimum requirements for milk productivity, live-weight and other characteristics. These are upgraded periodically with the trend of general improvement. Not all pure-bred animals are entered: the essential determinant being productivity, pure-bred animals of low productivity are not accepted.[4]

A dozen or so large research institutes and over fifty experimental stations, in addition to technical institutes and university faculties, cover the country. The entire field of research is guided in general terms by the V. I. Lenin All-Union Scientific Research Institute of Animal Breeding.

Great emphasis has been placed on artificial insemination in Soviet livestock breeding. I. I. Ivanov developed this technique with mares as early as 1897–1907, and the first large-scale experiment with the artificial insemination of ewes was carried out in Stavropol kray in 1928. By 1940 this

[1] Brezhnev and Minkevich (1958), 1961, 86–95.
[2] Burlakov (1961) 1966, 77, 79.
[3] Novikov and others (1950) 1960, 142, gave the number of state breeding grounds as 72, controlling over one million breeding cattle.
[4] Thus, Friesian (Holland) cows have to average 3·5 per cent. butterfat content unless of exceptional yield, in which case the minimum is 3·2 per cent.

technique was widely used for all classes of breeding livestock, though it was not widely used in other countries until after the Second World War. In 1960, about 70 million head of livestock were inseminated artificially compared with about 30 million in all other countries.[1] Higher percentages of dams were artificially inseminated in Denmark and in Czechoslovakia but total herds were much smaller in those countries. The use of sires lacking good characteristics has been greatly reduced, and the most productive strains of livestock developed and distributed in suitable regions throughout the country by this means.

Other experiments of zootechnical and veterinary kinds have been numerous, for example, a new method of castrating animals which, it is claimed, enables the animals to provide 10–12 per cent. more meat and wool by the age of one year and 15–20 per cent. after two years,[2] with some reduction in fat and increase in protein in the meat.[3]

Comparisons of numbers of livestock in different periods are here confined to the published statistics of the annual livestock census on the 1st January of each year. Numbers of livestock in the summer months would normally be much larger because of the number of young born in the spring,[4] but the winter figures have the advantage of indicating the size of the breeding herds and replacements. More detailed figures by classes and age-groups of animals are not available.

CATTLE[5]

Although outnumbered by sheep, cattle are the most important livestock in the U.S.S.R., in both output and fodder requirements. In 1968 they reached a peak of 97·2 million, including over 41 million cows.[6] These numbers were approxi-

[1] Milovanov (1960) 1964, 2–3.
[2] Baiburtsyan (1961) 1964, 2.
[3] Baiburtsyan (1961) 1964, 59.
[4] Durgin (1967) notes that a livestock count in January, 1938, showed 66·6 million sheep and goats in the U.S.S.R., whereas a count made in July of the same year showed 102·5 million. He points out that choosing a different census could result in one researcher reaching estimates 60 per cent. higher than another. This, however, is normal in agricultural statistics and the difference quoted presumably merely reflects a very moderate percentage of live births.
[5] *Krupnyy rogatyy skot,* lit. 'long-horned cattle' in Russian, as distinct from *skot,* which includes sheep, etc.
[6] Livestock numbers used are those quoted in regular Soviet statistical series as at 1st January each year.

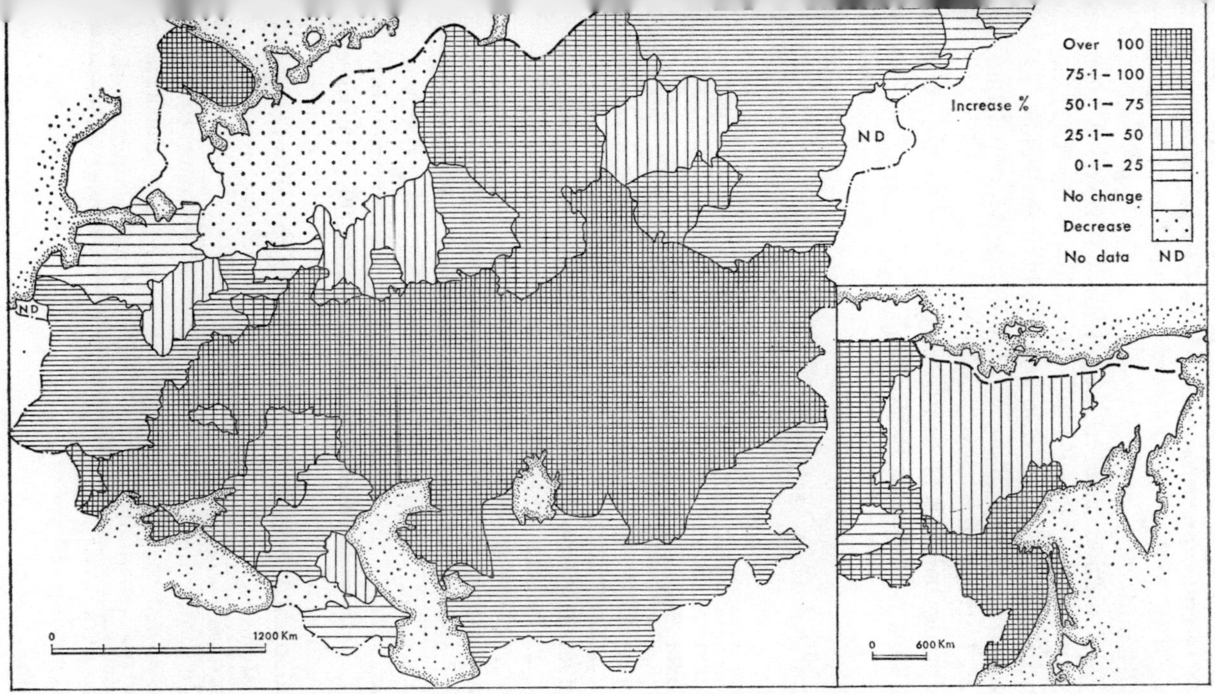

15. Cattle: percentage change in numbers, 1941–68

After recovery from wartime losses the expansion of cattle herds has been one of the most marked features of the agricultural development of the U.S.S.R. Only in the most northerly regions and in Georgia were cattle less numerous in 1968 than in 1941. The map shows the great belt of major development of cattle raising of the steppe and wooded steppe from the Ukraine to Kazakhstan and southern Siberia, with adjacent areas of west Siberia, Belorussia and Central Asia also recording substantial increases. Murmansk oblast and the Far East also show large percentage increases but the numbers involved are small mately double those at the end of the Second World War. On the eve of collectivization, as in 1916,[1] there were about 58 million cattle, and this number was again nearly reached in 1951 (57·1 million).

Of the 95·7 million head recorded on 1st January 1969 50 million were in the R.S.F.S.R., 20 million in the Ukraine, 7·5 million in Kazakhstan and nearly 5 million in Belorussia. Uzbekistan had nearly 3 million, and Lithuania, Azerbaydzhan, Georgia and Latvia had each over one million.

The changing balance of ownership of cattle by different types of holding is shown in Table 18.

TABLE 18

CATTLE BY TYPES OF HOLDING, U.S.S.R.
(millions)

	Total	*State farms*	*Collectives*	*Personal*
Total cattle 1951	57·0	4·2	28·0	24·8
Total cattle 1961	75·8	16·5	36·2	23·0
Total cattle 1969	95·7	28·6	39·8	27·3
Cows 1951	24·3	1·3	7·0	16·0
Cows 1961	34·8	5·7	12·8	16·3
Cows 1969	41·2	10·3	14·1	16·7

Sources: N.kh. SSSR 1965, 368; *N.kh. SSSR 1968*, 394.

The most striking feature of this table is the great increase in the numbers and percentages of the total national herd on state farms, and the slow rise in the numbers on collectives. In view, however, of the numbers of collectives converted into state farms, it is perhaps surprising that cattle on collectives have increased at all in recent years. Although the percentage of the total owned personally has declined, the numbers of cattle privately owned by the smaller farm labour force has actually increased slightly.

The table shows that in 1969 just under 30 per cent. of the cattle were owned by each of the groups made up by the state farms and the personal plot holders, leaving 41 per cent. in the collectives. With cows, however, 41 per cent. were in personal ownership, 34 per cent. in collectives and 25 per cent. in state farms.

[1] Soviet sources quote data for 1916 but not for earlier years, stating that figures for earlier years suffer from significant defects.

In the R.S.F.S.R., in 1969,[1] of the 49·7 million cattle of all kinds, collectives held 38 per cent., state farms 36 per cent., and private owners 26 per cent., i.e., the state farms had a larger share than in the U.S.S.R. as a whole, the kolkhoz share was slightly smaller and the personal ownership much smaller. There were about 21 million cows in the R.S.F.S.R., and, as in the U.S.S.R. as a whole, private ownership accounted for the highest proportion (37 per cent.), with collective and state farm ownership almost equal (32 and 31 per cent. respectively).

TABLE 19

OBJECTIVES IN YIELD OF MILK AND MEAT PER COW

Specialization type	*No. of cows in typical herds per cent.*	*Output*	
		*Meat kg. live-weight**	*Milk kg.†*
Dairy	60	130	3,000
Dairy-beef	50	240	2,400
Beef-dairy	40	300	1,500

* live-weight at slaughter.
† presumably, per lactation.

Source: Nemchinov (1956) 1960, 71.

The higher percentage of state farm ownership in the R.S.F.S.R. than in the U.S.S.R. as a whole is probably partly an expression of the importance of state farms in Siberia, and partly a reflection of the larger collective herds in other republics, especially Kazakhstan.

Soviet practice is to classify cattle into three main groups according to their purpose, or main purpose, viz. dairy, dairy-beef and beef-dairy. The differences in yield of milk and meat expected from cows of each group are indicated in Table 19.

These figures were given as targets for normal herds when only the best herds achieved this level. It is, of course, impossible to get really accurate yield figures, and great variation is concealed in milk yields because of variations in individual cows and different lactations as well as between different

[1] *N.kh. RSFSR 1968*, 239.

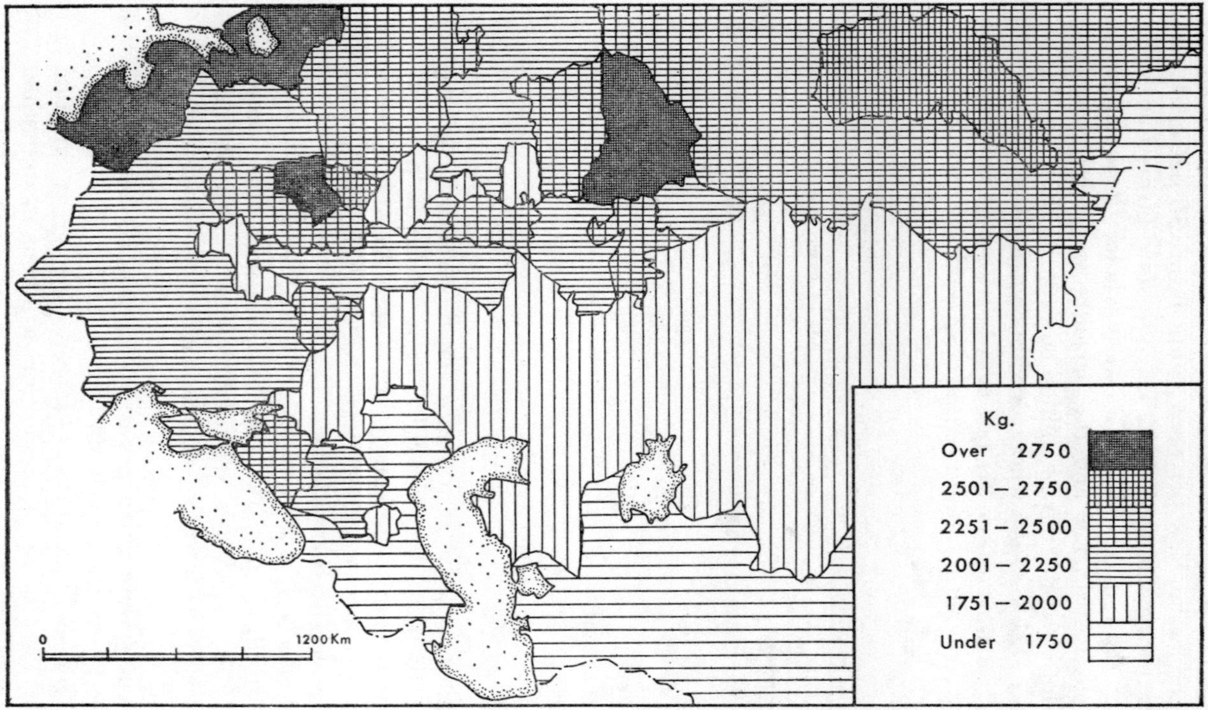

16. Milk yield: average per cow, 1965

Three areas traditionally associated with dairying stand out as continuing to lead the U.S.S.R. in milk yields—(a) the Baltic and Leningrad areas and the nearby Kostroma-Vladimir-Moscow area, (b) the lower Ural areas, notably Sverdlovsk oblast and (c) the Ob region of west-central Siberia. Most other western areas and the north Caucasus record averages of over 2,000 kg. per cow, but climatic conditions militate against high yields [illegible] areas, where the tendency has been to increase herds (Figure 15), largely for beef production breeds. However, by about 1965, yields in the better dairying areas were averaging 2,500–3,000 kgs. (Figure 16).

In 1969, when the average yield per cow was 2,232 kg. for all state and collective farms in the U.S.S.R., it was 3,184 in Estonia, 3,044 in Lithuania and 2,978 in Latvia and over 2,500 in the North-west and Central regions.[1] Average yields are double what they were in the long period of stagnation from 1945 to 1953 but are still well below figures achieved in western Europe. Averages are better on state farms than on collectives, but the gap is now smaller than in earlier years. At the end of the war the sovkhoz average yield was double that on the collectives, whereas it was only some 7 per cent. higher in 1968.[2]

The Soviet output of milk (including that used on farms) in 1968 was 82·1 million tons, the highest figure yet reported, 25 per cent. higher than three years earlier. Of the 1969 output, 45·8 million tons were produced in the R.S.F.S.R. and 17·8 million tons in the Ukraine. Most western areas, especially the Baltic republics, Belorussia, the Ukraine and adjacent regions of the R.S.F.S.R., have relatively high densities of cows and a surplus of milk products. Another area with some specialization in dairying occurs in western Siberia and north Kazakhstan.

While milk for consumption as liquid is the main object in the areas near the cities, processing for butter, cheese and milk powder absorbs most of the supply from farms distant from markets.

Butter production has been over one million tons per annum since 1965, whereas in 1961 it was only 781,000 tons and in 1946 only 186,000 tons. Earlier there were great fluctuations—falling from 129,000 tons in 1913 to 72,000 tons in 1932 during the collectivization campaign, rising to 226,000 by 1940 and falling again to 117,000 by 1945.[3] The progress since then has thus been the steadiest as well as the greatest in the Soviet butter industry. These figures, however, do not include butter produced in individual households, which would form much higher proportions of the total in the past

[1] *N.kh. SSSR 1968*, 409; *N.kh. RSFSR 1968*, 260.
[2] *N.kh. SSSR 1968*, 408.
[3] *N.kh. SSSR 1968*, 306.

than recently. Total butter exports in 1913 were 78,000 tons and, after falling almost to zero after collectivization, climbed back to 75,600 tons in 1968.[1]

As already noted, butter production is particularly well developed in the north-west forest zones and also in western Siberia, where the Kurgan area has long been particularly noted for butter, exports having been developed in the Tsarist period.

TABLE 20

USE OF MILK IN THE U.S.S.R.
(per cent. of total production)

Use	*Years*		
	1950	*1963*	*1970* (*plan*)
Whole milk	14·2	30·9	41·4
Butter	80·6	61·1	51·2
Cheese	3·9	4·0	6·2

Source: Markova and others (1967), 59.

While butterfat content and yield per lactation have been used as the main measures of dairy cow productivity in the Soviet Union, reliance on butterfat content has drawn some criticism in recent years. The Soviet Union has experienced the same tendency as other advanced economies for higher consumption of liquid milk, and present plans envisage a yet smaller percentage of milk being used for butter production.

It is therefore suggested[2] that payment should be by protein percentage rather than by butterfat, a problem already much discussed in other countries. The ranking of breeds and individual cows by protein content is different from that by butterfat. Simmenthal cattle in the Soviet Union, for example, with high butterfat records, show up less well in protein yield.

In a recent study of the Soviet dairy industry, Strauss concluded that the two main weaknesses in production were extravagant use of labour and poor feeding on both collective and state farms. Inadequate machinery, low quality of fodder and excessive numbers of inadequately fed stock, as well as unfavourable climatic conditions, contributed to this situation.[3]

[1] *N.kh. SSSR 1968*, 657.
[2] Markova and others (1967).
[3] Strauss (1970), 278.

Nevertheless, he pointed out that the Soviet population is now among the minority of peoples with an adequate total milk supply, though its utilization and presentation to the consumer need improvement.[1]

Meat production is also considerably higher than it was a few years ago. Figure 17 shows the regional improvements between 1960 and 1967. The 1968 total meat output of 11·6 million tons dead-weight[2] was 23 per cent. higher than the 1961–65 average, double the 1951–55 average and three times the 1946–50 average. This last was only 3·5 million tons, whereas the 1913 figure was put at 5 million tons.

Of the 1968 output, just over half (6 million tons) was produced in the R.S.F.S.R., and rather under one-quarter in the Ukraine.[3] Beef cattle are extremely widely distributed and are much more important than dairy cattle in the drier steppe regions, where fodder resources would be inadequate for dairying, and beef cattle range over large areas. Many of the steppe grasses and the hay made from them provide good feed for beef cattle, having a high protein content. Beef cattle are also more important in the wooded steppe than in the forest areas, where milk production is dominant. In the sugar-beet areas, the waste from processing provides additional fodder for beef production.

Despite the severity of the Russian winter it has been found practicable to keep cattle, including calves from birth, in unheated buildings. The Karavayevo breeding sovkhoz, specializing in the development of Kostroma cattle, found that although temperatures were below freezing, the young cattle remained healthy in an atmosphere in which humidity and evaporation of ammonia were kept low.[4] As early as possible in the spring the animals were taken out to pastures, notwithstanding the recurrence of frosts.

'Remote' livestock raising is practised in the south-eastern Transvolga steppes, Kazakhstan and the Transcaucasian republics. Beef cattle remain the whole year on pastures. In the semi-arid conditions the biggest problem is water supply. Natural reservoirs are rare and many are unsuitable

[1] Strauss (1970), 289.
[2] Soviet meat production statistics are quoted inclusive of fat and offals.
[3] *N.kh. SSSR 1968*, 401.
[4] Novikov and others (1950) 1960, 57.

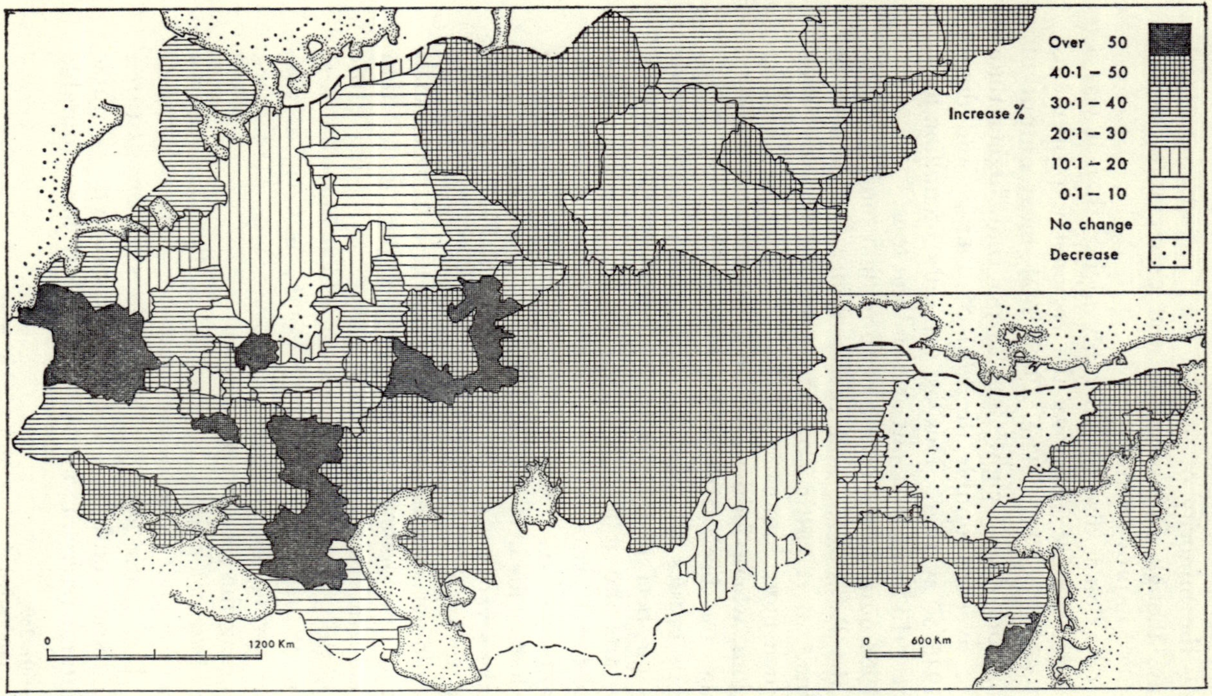

17. Meat output: percentage change, 1960–67

Belorussia and Lithuania, the lower Volga and north Caucasus and the southern Ural areas are outstanding in this map, with Kazakhstan, western Siberia and the southern Ukraine also showing large increases in output. Increases in cattle herds and improved fattening account for most of the improvement, with pig and poultry meat gaining in importance in some areas. Increased output of pigmeat was, however, largely at the expense of herd size which tended to diminish after rising in the early 1960s (Figure 14).

for use for cattle because of salt or infection. Cattle must be driven considerable distances from the grazing areas, wasting time and lowering the gain in weight, which ceases completely on hot summer days. Some pastures, distant from water, cannot be used at all. Lack of water is felt particularly in the late summer and even in the autumn because the natural reservoirs continue to dry out until then.

As a partial alternative to improving water supplies, supplementary feeding with melon crops (fodder pumpkins and water melons) has given good results. The Chkalov Scientific Institute of Cattle Breeding found that with adequate melon consumption (80–100 kg. per day) steers required no water at all and ate dry grass which otherwise they would not consume. It was found that 0·6 hectare of sown herbage and 0·1 hectare of melon crops per head enabled young steers to gain 172 kg. in the summer period.[1] The succession is from the dried-up natural pastures to sown grass and use of melons and pumpkins, minimizing the use of concentrates.

Tallow, suet and hides are by-products which attain greatest importance in the driest areas where milk and meat production is lowest. Store cattle are also sent from these areas for fattening in the richer zones.

Fattening on waste products is one of the methods used to supplement pasture and grain resources in livestock rearing. *Zhom* and *barda*, waste products from the sugar beet and alcohol industries respectively, are particularly important. Even before the war these two provided 15–20 million centners of feed units, sufficient for fattening 2 million cattle. To avoid waste, fattening was in some cases specially organized near factories producing sugar or vodka. Techniques of processing and conserving these fodders, such as drying, freezing and ensiling have been more developed recently.

CATTLE BREEDS

In pre-revolutionary Russia, attempts at breeding cattle scientifically were on a small scale compared with developments in western Europe and the British Isles. In the south of Russia, where cultivation was relatively well developed, cattle were reared mainly for draught purposes, in the northern

[1] Novikov and others (1950) 1960, 99, 354.

and central provinces largely to yield dung for fertilizing. Stimulus to improve the output of cattle came with the rapid growth of industrial towns in the eighteenth and nineteenth centuries. In particular, the rapid growth of St. Petersburg from its foundation in 1703 provided a good market for milk, butter and cheese from the northern provinces, while Moscow, though suffering a temporary reduction in population, numbered about 150,000 people and resumed its growth to about 175,000 in the early 1790s. Smaller, but locally important markets were offered by the expanding textile towns and other industrial centres.

The cattle of this area, known as Great Russian or Forest cattle, were generally small and of varying type, though some had the reputation of good breeding cows with high milking qualities. This applied to the meadowlands of the river valleys of the Northern Dvina, Vychegda, Kostroma, Oka, Klyazma and other tributaries of the Volga, which were also relatively well placed to exploit the city markets.

Conscious efforts to improve these Russian native cattle began on a small scale in the eighteenth century. The cattle of the Northern Dvina valley around Kholmogor, which enjoyed the reputation of being unusually well fed and consequently productive, attracted the attention of Peter the Great. On his initiative Dutch cattle were imported through the port of Arkhangelsk. In 1765 Catherine II ordered 6 bulls and 36 cows from Holstein for the Arkhangelsk and Vologda provinces and further imports were made at intervals throughout the nineteenth century.[1]

Growth of markets similarly stimulated the growth of dairy industries in other parts of the country. The development of metallurgical and other industries in the Urals provided markets for the output from the improved Tagil cattle, in the evolution of which Kholmogor, Dutch and other native and imported breeds were utilized from the late eighteenth century.

As in other countries, the improving movement could be initiated and sustained only by those with the resources to experiment and, in this case especially, the wealth to import livestock from abroad. Besides the royal family, nobles and

[1] Novikov and others (1950) 1960, 144.

landed gentry, the religious houses played a part as they had earlier in western Europe. A noteworthy contribution was made in the nineteenth century by the Nikola-Babayev monastery, which brought Algau cattle to the Kostroma district to help with the improvement of the local cattle.

Sheremetev's estate at Gorbatov, near Nizhniy Novgorod (now Gorkiy), was the scene in the nineteenth century of improvements to local breeds through the medium of imported Red Tyrolean cattle. Sheremetev was particularly interested in cattle breeding to supply his tanneries with hides, but the result was the emergence of the Gorbatov Red dual-purpose cattle, now one of the basic dairy breeds in the Soviet Union. The Brown Swiss breed was widely imported in the nineteenth century for its dual-purpose qualities, and Switzerland also provided the Simmenthal, which became, with its cross-bred derivatives, one of the most important dual-purpose breeds in Russia. Simmenthals were shown at the 1869 exhibition at St. Petersburg, and accounted for 47 per cent. of the cattle shown at the exhibitions 1907–09.[1]

Financial stringency hindered the spread of the improved cattle. Landowners commonly allowed their bulls to run with the cows belonging to the peasants, but this, in itself, resulted in the dilution of improved strains, and few peasants could afford the better stock. The improved cattle also needed good feeding if they were not to revert and some, such as the large, heavy Simmenthals, could not be adequately supplied from the small holdings, most of which were still worked on the three-field system and lacked provision for fodder crops.[2] Simmenthals were adopted by peasants mainly where the reforms had created *otrub* and *khutor* (separate) holdings.[3] No doubt these reasons as well as increased availability of locally improved stock led to increased attention by the zemstvos to less-demanding native breeds between 1909 and 1914. The zemstvos were active in promoting local dairy industries. The Vyatka provincial zemstvo organized a cheese

[1] Novikov and others (1950) 1960, 301.

[2] In some areas, as in the Ukraine, leguminous fodder crops and roots were grown for cattle feed early in the nineteenth century and the sugar manufacturers of the beet-growing areas had by-products available as fodder and so stimulated cattle raising in the southern areas.

[3] Novikov and others (1950) 1960, 301.

factory in Istobensk in 1870, and a butter factory in 1895. Between 1895 and 1915 more factories were opened until in 1915 there were about fifteen and a butter manufacturing artel was set up, with 114 members owning cows. Working as co-operatives many such artels helped in breed improvement. The Istoben breed of dairy cattle was developed in the area during this period and a control union was established in 1914.[1]

The zemstvos could not get the research support they needed because agricultural institutes were so few and those that did exist had to deploy their resources over too wide a field. Crop husbandry commanded greater attention than livestock rearing. Nevertheless, the institutes and independent bodies, such as the Free Economic Society, sponsored investigations into breeding and rearing cattle and published reports which encouraged the formation of breed control societies. These arranged exhibitions, competitions, purchase of valuable bulls and other measures to improve breeds. The Moscow Agricultural Institute helped introduce improved breeds to surrounding areas, as in placing four Brown Swiss bulls into the Kostroma district in 1898.[2] Simmenthal, Bestuzhev, Tagil and Shorthorn are among the breeds that have subsequently contributed to the emergence of the Kostroma as a breed well adapted to the long winters and relatively humid conditions of the central R.S.F.S.R.

Whereas conditions in the northern and central Russian provinces were not too different from those in western Europe for successful breed adaptation, the extremes of climate in other regions of the U.S.S.R. militated against their success, and much of the effort in the Soviet period in cattle breeding has been aimed at producing suitable breeds by crossing imported and native stock. Thus Friesian, Brown Swiss and, especially, Simmenthal cattle found difficulty in surviving in Central Asia and Transcaucasia, but cross-breeding with local cattle has produced new breeds which are well adapted. Crossing Kazakh and Astrakhan cattle with Herefords produced the Kazakh Whitehead which has a high live-weight, matures quickly and gives a high yield of milk.

[1] Novikov and others (1950) 1960, 245.
[2] Novikov and others (1950) 1960, 144.

Various breeds of cows have been crossed with Jersey bulls to improve butterfat content of the milk. By this process the Institute of Genetics of the U.S.S.R. Academy of Sciences raised cows which from their first calving yielded in six months in 1957 an average of 2,179 kg. of milk with an average butterfat content of 5·2 per cent. Butterfat contents up to 6·1 per cent. were recorded with second calvings.[1] High butterfat content is a feature also of the dual-purpose Alatau breed, which has been evolved to suit the conditions on mountain grazings, using the techniques of absorptive cross-breeding and selection over a long period. The mountain pastures have also been improved and more nutritive feeding and stall shelter provided during the winter. This breed has been supplied to the Uzbek and Turkmen republics and also to China, Korea and India.[2] Considerable success has been achieved in meeting the needs of extremely hot and mountainous regions by means of hybridization of cattle (*Bos taurus*) with other species, notably zebu and yak.[3]

Zebu cattle have played an important role in the southern regions of the U.S.S.R. The zebu has many faults, including low yields and low butterfat content of milk, slow growth and light finished weight. It is, however, well adapted to hot climates and has not only responded well to improvement but has itself proved a valuable improver breed. Milk yields of Azerbaydzhan zebus have been improved by careful treatment of the udder, liberal feeding and selection without infusing blood from other breeds. Yields rose to 1,620 kg. per lactation with 5·1 per cent. butterfat, and individually to 2,000 kg. per lactation, compared with an average of 300–400 kg. Further improvement was effected by cross-breeding. First-calf heifers from zebu cows and Red Steppe bulls gave 2,480 kg. in a 300-day lactation with 4·8 per cent. butterfat, and weighed 372–394 kg. compared with the 220 kg. of typical unimproved zebus.[4] Red Steppe cattle also give milk of low butterfat content so that the result of the cross is particularly

[1] Brezhnev and Minkevich (1958) 1961, 89–90.

[2] Vsyakikh (1961) 1966, 93.

[3] Hybridization means the mating of animals of different species, whereas crossing denotes mating of different breeds of the same species. The different methods used in forms of crossing as well as hydridization, are discussed by Novikov, Startsev and Arzumanyan (1950) 1960, 99–129.

[4] Dobrynin (nd, 1964?), 92–93.

interesting. The introduction of a zebu bull resulted in improvement of the constitution as well as the butterfat yield of the Red Steppe cows.[1] Zebus have also been crossed in the U.S.S.R. with Angus, Hereford, Shorthorn, Charolais, Swiss, Simmenthal and the Soviet Lebedinskaya breeds.[2]

The yak is raised in the mountainous parts of Kirgizia and Tadzhikistan as well as in Mongolia and Tibet. Unlike the zebu it is adapted to cold conditions and is small, but can exist with little food. Various crosses have been tried, with sterility occurring in some of the hybrids, but not in the cows of the zebu x yak.

The dairy buffalo is important in Azerbaydzhan, and also in Georgia, Dagestan, Armenia and the north Caucasus region. Able to withstand high temperatures provided it has water in which to bathe, and possessing a strong constitution, it is exceptionally useful as a working animal. It also yields milk with butterfat content up to 8·2 per cent. and milk production costs are among the lowest in the U.S.S.R. Good leather is obtainable from its hide. Hence, development of this breed aims at offsetting faults of late maturing, poor quality meat and low volume of milk yields.[3]

SHEEP

In most parts of the Soviet Union, sheep are much less important than cattle. Mutton, however, is generally well liked, especially in Central Asia and the Caucasus, and there is a high demand for sheepskins in addition to the basic industrial market for wool. The pelts of Karakul lambs are a traditional export and there are large areas of arid or mountain terrain which can be exploited better by sheep grazing than by other enterprises.

There have been greater absolute fluctuations in sheep numbers than in any other class of livestock. From nearly 90 million in 1916 numbers fell sharply in the civil war, recovered and then were more than halved during the collectivization drive. A recovery to 80 million occurred by 1941,

[1] Dobrynin, 103–104.
[2] Arzumanyan (1961) 1966, and Dobrynin, 91.
[3] Agabeili (1961) 1966; Arzumanyan (1961) 1966.

only for war to produce a fall to below 60 million. Since 1948 there has been much steadier progress and on 1st January 1969 the enumeration showed the record total of 140·6 million in the whole of the U.S.S.R.[1] About 21 per cent. of the total are privately owned, compared with 15 per cent. in 1951 but 33 per cent. in 1941. The increase in sheep numbers since 1941 has thus been achieved mainly in the state and collective farms, but the personal ownership contribution has been permitted to increase after the low level to which it had been forced.

About 44 per cent. of the total sheep are in the R.S.F.S.R., and 24 per cent. in Kazakhstan, with some 6 per cent. in each of the Ukrainian, Kirgiz, and Uzbek republics. Relatively high densities are found in the latter two areas and in Transcaucasia and Moldavia, while in the R.S.F.S.R. there are marked concentrations in the north Caucasus, lower Volga and Don regions and the Mordov, Chuvash, Tatar, Bashkir, Tuvinian and Buryat A.S.S.Rs. (Figure 18).

In the areas where seasonal droughts or snow cover make certain types of pasture inaccessible, seasonal nomadism or transhumance is practised. This is particularly true of Central Asia and the Caucasian regions, where mountain pastures are relatively close to lowlands affected by dry conditions in summer but not by particularly heavy snowfall in winter, so facilitating interchange.

The average yield of wool, the main object of sheep rearing in the Soviet Union, has improved over the years, but only slowly. It was returned as 2·5 kg. per sheep in 1940 and, after the wartime dislocation, did not return to this figure until 1955. Since then the average for collective and state farms has not fallen below this figure, but not until 1967 was 3 kg. claimed.[2] The average for collective and state farms in the R.S.F.S.R., was 3·5 kg. in this year,[3] reflecting the generally heavier weight of wool obtained from the specialist sheep-breeding farms of the North Caucasus and Volga regions. The highest regional average is consistently recorded by Stavropol kray, 4·9 kg. in 1968, followed by Rostov oblast, 4·3, and then

[1] *N.kh. SSSR 1968,* 393.
[2] *N.kh. SSSR 1968,* 410.
[3] *N.kh. RSFSR 1968,* 262.

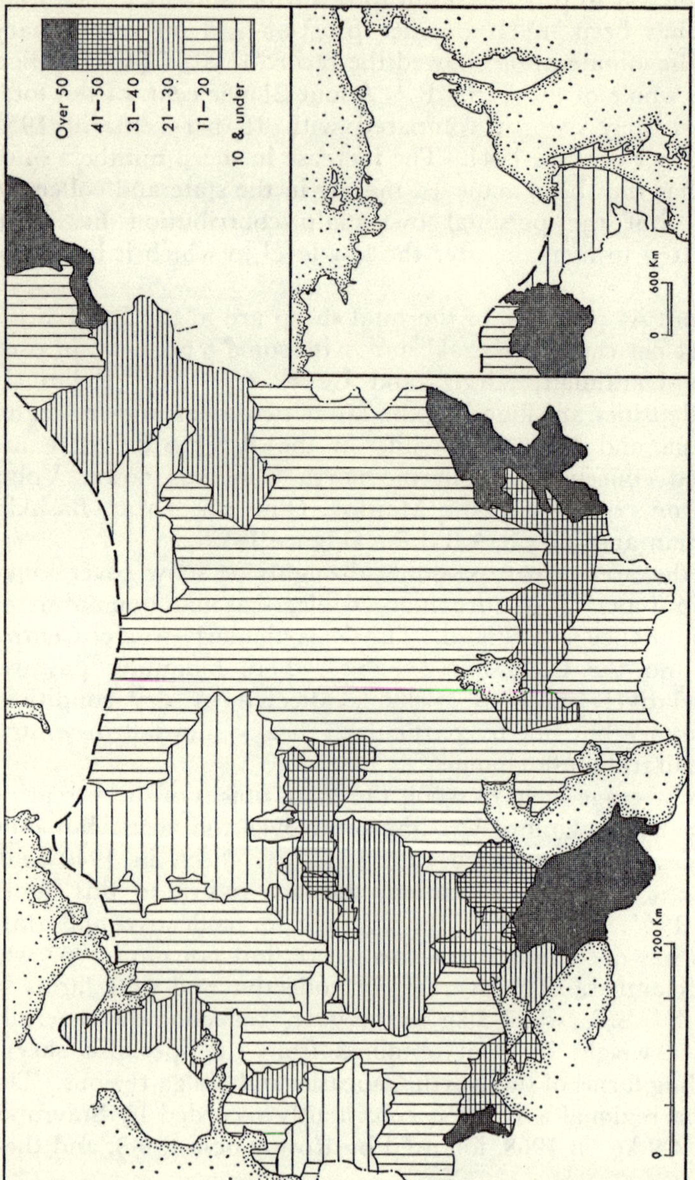

18. Sheep per 100 hectares total area, 1966

the Kalmyk A.S.S.R., 4·2 kg. and Astrakhan oblast, 3·9 kg.—a fall from 4·2 kg. in the two previous years.

As a result of the increased numbers of sheep and improved yields, state purchases of wool have increased substantially in recent years. In 1966 they regained the previous highest level of 380,000 tons achieved in 1963, and in 1968 rose to 428,800 tons, of which 201,700 tons came from the R.S.F.S.R. and 107,200 tons from Kazakhstan.[1]

TABLE 21

RELATIVE VALUES OF PRODUCTS FROM SHEEP

Type of sheep	*Per cent. of Total Value*		
	Meat	*Wool*	*Sheepskin*
Meat-fat (lard)	80	15	5
Meat-wool	75	20	5
Coarse wool	65	25	10
Fine fleece	40	50	10
Astrakhan (Karakul)	20	20	60
'Fur'	67	15	18

Sources: compiled from Nemchinov (1956) 1960, 72 and other sources.

Output of meat is less easily calculated than wool production because much of it does not pass through the hands of the state purchasers. Factory output of meat and meat products from sheep was about 400,000 tons annually through the 'sixties, compared with 169,000 tons in 1913. At that time mutton constituted over one-tenth of all meat production, but in 1968 the 412,000 tons from sheep constituted only 6 per cent. of total output. It was then exceeded in factory throughput over four times by pigmeats and nearly nine times by beef and veal.[2] Total output of mutton, with goatmeat added in, was estimated as one million tons in 1968.[3]

Based on state prices, the relative values of the products from sheep are approximately as given in Table 21.

Considerable efforts have been made to increase productivity of sheep in recent years, reasonable output levels which

[1] *N.kh. SSSR 1968*, 407.
[2] *N.kh. SSSR 1968*, 304.
[3] *N.kh. SSSR 1968*, 320.

might form objectives in rearing having been stated as follows:

TABLE 22

OBJECTIVES FOR OUTPUTS IN SHEEP BREEDING

Specialization type	*No. of ewes in typical flocks per cent.*	*Output per Ewe*	
		Meat (kg. live-weight)	*Wool (kg.)*
Meat-fat (lard)	50	65	3·8
Meat-wool	70	60	4·5
Fine fleece	60	40	6·6
Semi-coarse wool	70	52	4·7
Astrakhan (Karakul)	75	7	2·5
'Fur' type	60	54	2·0

Source: Table compiled by V. F. Chervinskiy, quoted by Nemchinov (1956) 1960, 72.

SHEEP BREEDING

In pre-revolutionary Russia sheep were mostly of coarse-woolled varieties. The emphasis in breeding during the last few decades has been on developing fine-woolled sheep, with the All-Union Scientific Research Institute of Sheep and Goat Breeding and nearby farms in the Stavropol area leading the work. Particularly well known are the Bolshevik and Soviet Fleece state farms, the latter being the home of a new fine-woolled breed.

Emphasis has been placed on improving the Merino, the most numerous breed in the Soviet Union. Various measures have been found to be beneficial, including early lambing, which for the north Caucasus area means in February or early March, not using ewes younger than two years for breeding, and using unheated buildings for lambing to favour production of hardy and vigorous lambs.[1] In addition, of course, rams of proven productivity, use of artificial insemination to make maximum use of selected sires, and good rations of nutritious fodder are essential. The latter includes sowing of legumes in pastures for summer use and provision of silage in winter. Improved feeding alone has been claimed to have

[1] Brezhnev and Minkevich (1958) 1961, 88–89.

raised live-weight of yearling ewes from 41 to over 55 kg. with wool clip rising from 5·43 to 8·23 kg. in the same groups.[1]

The Caucasian Merinos crossed with Askanian Rambouillet rams has produced an improved fine-woolled breed called the Caucasian Rambouillet. Similar improvement has been achieved with the Dagestan, Altay, Kirgiz and other mountain breeds, giving rises from 1·92 kg. of coarse wool to 3·36 and 3·64 kg. of medium wool respectively. Medium-woolled sheep have also been bred to suit the harsh conditions of northern areas in Siberia, with a wool clip of over 5 kg. from ewes and 8–10 kg. from rams.[2] Other improved breeds include the Kazakh fine-woolled, Azerbaydzhan Mountain Merino, Georgian, Gorkiy, Kuybyshev and Salsk.

In the better managed flocks a lambing percentage of 140–150 is frequently achieved, but the average even in the more favourable areas is much lower. High averages, however, are attained in some of the Karakul flocks, kept mainly in the Central Asian desert regions, rising to over 200 per cent., though few normally exceed 150 per cent.

The record for regularly high lambing rates would, however, seem to be held by the Romanov sheep of the Yaroslavl region. Over 250 per cent. is claimed here, and some ewes have delivered five lambs. The high fertility is also reflected in the frequency with which these ewes lamb twice in one year. Because of this feature lambing percentages vary widely, an example being 1962–65; 257, 99, 307, 200, average 216 per cent.[3] In addition, the Romanov may yield 60–80 kg. of meat, and 4·0–4·5 kg. of wool.

In the severer climates, losses of lambs are high, but survival rates may be improved by earlier, rather than later, lambing. Thus, collective farms of Kizhinga rayon, Buryat-Mongolian A.S.S.R., reported that, with 80 per cent. of the lambs born in the winter, only 4 per cent. were lost, compared with 15 per cent. of those born in March and 22·5 per cent. of the April-born lambs.[4]

Artificial insemination has been particularly valuable in sheep breeding. In 1959 nearly 33 million ewes, 61 per cent.

[1] Brezhnev and Minkevich (1958) 1961, 89.
[2] Brezhnev and Minkevich (1958) 1961, 92–93.
[3] Florenskiy (1967).
[4] Ignatov (1956) 1960, 245.

of the total, were artificially inseminated.[1] As early as 1936 the semen of one ram was used to inseminate 15,000 ewes in one season and since then this has been considerably increased, 21,854 being recorded in 1958,[2] compared with using about 60 rams per 1,000 ewes in natural mating. The possibilities of flock improvement are thus immense.

There has been considerable criticism of development of intensive sheep breeding in the more fertile areas, on the assumption that this is a branch of husbandry advantageous only for exploiting areas of poor land. This is partly because of the emphasis on wool and neglect of meat production. A typical criticism of the introduction of sheep into what is considered cattle country is the comment 'it led, as a rule, to an intolerably high cost of wool', state farms in Vologda and Novgorod oblasts being named as having produced wool at a cost several times greater than the state purchasing price.[3]

It has, nevertheless, been recognized that sheep are not to be judged wholly as competing with cattle, being good gleaners, utilizing fodder left after cattle grazing, and it has been suggested that two sheep can be maintained per cow on cattle pastures.[4] In demands on fodder seven sheep are equated with a cow, but labour requirements are low, one cow being equated with 15–20 sheep.[5]

In the semi-desert and mountainous areas sheep should play a large part, but their potential is far from fully exploited. Sheep could certainly be bred on some of the almost unused pastures of Kazakhstan, said to amount to over one hundred million hectares. The hardiness of wethers in these conditions of poor pastures and severe winters commends them for special attention, and, according to the Muyunkum state farm for sheep breeding in the Kazakh republic, they could be used for a stable and profitable undertaking.[6]

GOATS

Goats were very popular among both collective farms and private plot holders in the late 'forties and early 'fifties, rising

[1] Milovanov (ed.), (1960) 1964, 4.
[2] Milovanov (ed.) (1960) 1964, 10.
[3] Novikov (1956) 1960, 286.
[4] Nemchinov (1956) 1960, 72.
[5] Zaltsman (1956) 1960, 231.
[6] Nemchinov (1956) 1960, 72.

to a peak of 17·1 million in the U.S.S.R. as a whole in 1952. Over half were privately owned. Numbers then declined, and an increasing proportion of the remainder were enumerated as personal livestock. By 1960 they had fallen to 7·9 million, over 80 per cent. personally owned, and by 1965 to 5·5 million, at which they have remained roughly constant.[1]

While nowhere in the Soviet Union are goats of great importance, they have some significance in the areas where sheep are relatively unimportant. Thus, in 1966, the ratio of goats to sheep was 1:6 in the far-apart Tadzhik republic and North-west region of the R.S.F.S.R., 1:8 in the Volga-Vyatka region, 1:9 in the Far East, 1:10 in the Central region, 1:11 in the Central Chernozem region and 1:12 in the Ural region and in the Ukraine. At the other extreme, the ratios were 1:40 in East and West Siberia, 1:44 in Kirgizia, 1:55 in the North Caucasus and 1:61 in Kazakhstan. The R.S.F.S.R. average was 1:21 and the Central Asian and Transcaucasian average 1:20.

PIGS

Pigs are less important in the Soviet Union than either cattle or sheep in numbers and fodder requirements, but their contribution to the meat supply is more than four times as great as that of sheep. From 1953 to 1966 pigmeats were estimated to exceed beef and veal in total output by dead-weight in every year except 1964. In factory output, however, pigmeat products were less than products from cattle.[2]

There has been considerable fluctuation in the numbers of pigs, roughly conforming with those of cattle and sheep, but not as marked a 'pig cycle' as in many western countries. After the fall in the collectivization drive, numbers recovered more quickly than with cattle and sheep, and between 1936 and the war generally exceeded the 1916 figure of 23 million. They were halved during the war but reached 24·4 million in 1951 and, with minor fluctuations, rose to 70 million in 1963. (Figure 14). The grains and fodder crisis of that year led to a reduction to 40·9 million, then there was a recovery to 59·6 million in 1966, followed by a small decrease during

[1] *N.kh. SSSR 1968*, 393.
[2] *N.kh. SSSR 1968*, 320, 304.

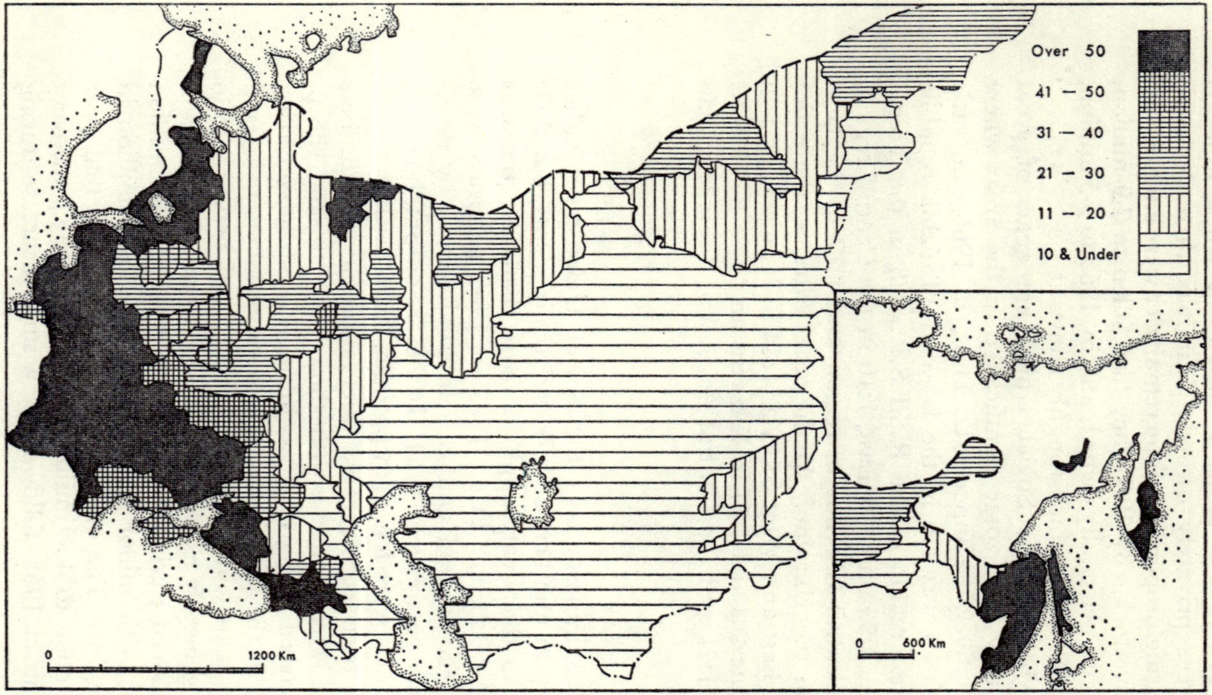

19. Pigs per 100 hectares of arable land, 1966

The importance of the Ukraine, Belorussia, the Baltic states and adjacent areas of Russia in the rearing of pigs is seen in this map. High densities of pigs are also recorded in the Far East but the numbers are small, as is the amount of arable land. Pastures and forests provide food for pigs but for the most part they are kept on the produce of arable land and hence are here plotted in relation to it. The contrast between the numbers of pigs on arable land in the west and in Central Asian areas is noteworthy. Traditional attitudes among Moslem [illegible] may still be relevant, though the numbers of pigs in all Central Asian republics and Kazakhstan

that year and a sharper drop in numbers during 1967 to result in a total of 50·8 million in 1968 and 49 million in 1969.

The versatility of pigs enables them to be reared on a wide range of fodders and types of terrain, but they are associated in the Soviet Union particularly with the grain, potato and sugar-beet areas. Highest densities are, therefore, in the Ukraine, Moldavia, Belorussia and the Baltic republics, especially Lithuania (Figure 19). In the R.S.F.S.R., which has over half the total pigs, the main areas of concentration are close to those already named, the central black-earth and north Caucasus areas having the highest concentrations, with a thinning-out of numbers in the Volga region and eastwards. High densities in relation to arable land are recorded in the Far East but the numbers are small.

In the U.S.S.R. as a whole, the great increase in pig numbers up to 1966 was shared by all sectors, but not proportionately, the state farms gaining most. The proportion of privately owned pigs fell from 36 per cent. in 1951 to 26 per cent. in 1969. The private sector in recent years has maintained about the same numbers as the state farms, with considerably more on the collectives.[1]

In spite of decreases in numbers of pigs on personal plots at times of crisis, the plot holder has succeeded to a remarkable extent in keeping his livestock through difficult periods with his own carefully attended fodder crops, skim milk and domestic waste. There has also been response to high prices paid for pigmeats, particularly since 1965. Raup may well be correct in suggesting that the U.S.S.R. may now have reached the stage in its development at which it is capable of generating a true price-induced pig cycle, rooted in the traditional lag between price signals and production responses,[2] and aided, of course, by weather conditions and consequent fluctuations in harvests.

Pigs are efficient converters of industrial waste, such as that from sugar-beet processing, domestic scraps and a wide variety of fodders, fresh, ensiled and concentrated, while the rearing and fattening periods are short. Part of the attraction to Khrushchev of the maize programme was the hope of

[1] *N.kh. SSSR 1968*, 394.
[2] Raup (1967), 260.

creating a pig industry on a scale and with an efficiency akin to that of the American corn belt. Self-feeding is widely used for pigs, with concentrates, silage and root crops, while feeding frames are recommended for use when pasturing pigs.[1] Potato fields provide good feed for fattening.

In the past, pigs were valued for their output of fat, which was usually scarce in most of Russia, thus encouraging production of heavy sows and large litters. The current practice in most western countries of aiming for pigs of bacon weight and generally encouraging quality rather than quantity has been slow to penetrate to the Soviet Union, where fat meats in general remain more acceptable than in western countries. Pigs are still classed as either lard, lard-meat or meat pigs, but the emphasis is now on the meat. The bacon type has been emphasized in recent breeding, with considerable use of the Large White for crossing with native breeds, for example, the Breytov breed in the Yaroslavl area and the North Siberian breed. On the other hand, measured feeding has been claimed to give better results with Belorussian Black-pied hogs than with Large Whites. The Grey Omsk breed has been developed to suit harsher conditions.[2]

Artificial insemination was introduced in actual productive pig breeding in 1959 after a long period of experiment.[3]

POULTRY

Nowhere is the recent advance of the state sector in Soviet agriculture more evident than in poultry raising. As recently as 1960, over 80 per cent. of the eggs produced in the U.S.S.R., came from the privately owned flocks. By 1968, the percentage was down to 60.[4] In terms of eggs marketed the percentage produced by collective and state farms rose from 46 in 1960 to 76 in 1968. The contribution of the state farms rose even more dramatically in this time from 21 per cent. in 1960 to 50 per cent. in 1968 of all eggs marketed.[5]

Holders of personal plots are not restricted in the numbers

[1] Brezhnev and Minkevich (1958) 1961, 86–89.
[2] Brezhnev and Minkevich (1958) 1961, 86–89.
[3] Milovanov (ed.), (1960) 1964, 4.
[4] *N.kh. SSSR 1968*, 321.
[5] *N.kh. SSSR 1968*, 323.

of poultry they can keep, and poultry suit small-scale enterprises It has, however, been state policy to develop large, modern poultry farms for intensive production of poultry meat as well as of eggs.

The number of poultry in the U.S.S.R., rose from 293 million in 1951 to 516 million in 1961. After subsequent fluctuations the number in 1969 was 543·5 million. About 55 per cent. are in the R.S.F.S.R. and 25 per cent. in the Ukraine.[1]

Although poultry are widely distributed, heavier densities have been traditionally associated with the grain-growing areas, notably the Ukraine and adjacent areas of Moldavia and the Kuban. With the growth of specialized poultry farms, however, nearness to the market has become more important than proximity to fodder supplies, which can be transported in bulk. The large, new farms are therefore located near Moscow and other large cities, in some cases with central incubating stations producing chicks for distribution to the farms.

TABLE 23

TYPES OF POULTRY ON COLLECTIVE FARMS IN THE R.S.F.S.R.
(millions, at beginning of year)

	1941	*1951*	*1961*	*1966*
Total poultry on collectives	15·3	32·0	41·9	30·5
of which, hens	14·6	29·8	32·8	27·3
geese	0·5	1·6	0·2	0·1
ducks	0·2	0·4	4·1	0·4
turkeys	0·0	0·2	0·1	0·1

Source: N.kh. RSFSR 1965, 291.

Poultry meat, though long occupying a very minor position in the Soviet Union, with production generally about half that of mutton, has recently risen in importance and in 1967 was estimated at 800,000 tons—equal to 80 per cent. of the mutton output (Figure 14).

Ordinary fowl account for all but a small percentage of poultry. Table 23 indicates the position on collective farms in the R.S.F.S.R.

[1] *N.kh. SSSR 1968,* 398.

HORSES & OTHER DRAUGHT ANIMALS

Draught animals in the Soviet Union include horses, oxen and cows, asses, mules and donkeys, camels, water buffalo, yaks and reindeer. While the last four are used only in the regions to which they are especially adapted, the others are widespread, though their numbers are much reduced from those recorded before the development of mechanization.

The declining percentage of total power used in agriculture derived from working animals, according to official calculations, is given in Table 24.

TABLE 24

ESTIMATED CONTRIBUTION OF WORKING ANIMALS TO TOTAL POWER USED IN AGRICULTURE IN THE U.S.S.R.

At end of year	*million h.p. (mech. equivalent)*	*Per cent. of total*
1916	23·7	99
1928	—	95
1940	10·6	22
1950	7·3	12
1960	4·7	3
1965	3·7	1·5
1968	3·9	1·4

Sources: N.kh. SSSR (1956), 150; *N.kh. SSSR 1968*, 413.

The number of horses declined from 38 million in 1916 to 21 million in 1941, 14 million in 1951, 10 million in 1961 and 8 million in 1966, since when numbers have been stable.[1] Many of the horses still kept are used mainly for sporting and recreational purposes, though large numbers still work in agriculture.

Horses have particular value for haulage when unsealed roads are at their worst—often impassable—in the spring and autumn. They also retain special usefulness in the mountain areas, and here and in the pastoral regions of eastern Siberia and Kazakhstan they are still important in livestock-rearing activities.

European and Arabian strains are valued in breeding, which has been improved considerably during the Soviet period. The steppes of Cossack settlement and many areas

[1] *N.kh. SSSR 1968*, 393.

in Central Asia and Siberia, where small, hardy Mongolian breeds are popular, still show something of the traditional interest of the peoples of these regions in horses.

REINDEER

In the tundra and the northern tayga reindeer remain the only important livestock. They provide meat, skins, and means of transport, hauling sledges or being ridden or used as pack-animals.

The numbers of reindeer can hardly be quoted accurately, but the official records show an increase from about 2 million in most years of the 1941–61 period to nearly 2·5 million in 1969. In 1961 20 per cent. were included in state farms, 14 per cent. were in personal ownership and the remainder in collectives. In 1969 the number in personal ownership had risen slightly but then formed only 13 per cent. of the total, while the numbers in state farms had risen to 60 per cent.[1]

The main increases in numbers have been in the Far Eastern region, especially the Chukot national okrug and other parts of Magadan oblast, which contains half of the reindeer of the Far East region. There is a broad spread of reindeer herding across the Koryak national okrug (Kamchatka oblast) and the Yakut A.S.S.R., with small numbers as far south as Chita oblast and the Buryat A.S.S.R. There are considerable concentrations in the Yamalo-Nenetskiy national okrug (Tyumen oblast), the Nenets national okrug (Arkhangelsk oblast), the Komi A.S.S.R., and Murmansk oblast.[2]

Herds are grouped roughly into domestic and wild. There is no great difference, but the domestic animals are the ones trained for work and for them supplementary fodder is now frequently provided in winter, being transported in for the purpose. Other than this, feeding is from natural pasture, the animals grubbing under the snow for lichens and mosses. There is still migration between the moss-grazing areas of the northern tayga and wooded tundra in the winter and the true tundra and coast, where insect pests are less troublesome in summer, but the continuous trekking of the past has been largely eliminated by planned grazing. Pastures require

[1] *N.kh. RSFSR 1968*, 249.
[2] *N.kh. RSFSR 1968*, 249.

up to ten years to regenerate and large areas are required accordingly for each animal. Herders can care for several hundred animals each, particularly since the introduction of aircraft, including helicopters, which have proved very successful in these regions. Grazing control, disease prevention and protection from predators have all helped in the improvement of productivity of herds.

Reindeer herding is combined with other economic activities, notably hunting, sealing and fishing, according to the resources of the area. Even in the Chukot area, however, reindeer are said to account for about half the total earnings and to provide two-thirds of the trade in meat, after an increase in numbers by 48 per cent. during the Seven Year Plan. The utilization of pastures, especially distant and little exploited areas, has been much improved. The settlement of the population in permanent villages was speeded up during the early 1960s with electrification and provision of basic amenities in all central settlements of the collective farms.[1]

[1] Gradov (1967), also *Izvestiya*, 18.12.66; *C.D.S.P.* 19, (4) 15.2.67, 14.

CHAPTER 10

Regions of Agricultural Specialization

As a result of the great size and physical diversity of the Soviet Union a wide range of combinations of crops and livestock can be raised within its territory. These can be combined in many different ways, responding to national and local needs as expressed in plan requirements and prices as well as the physical limits of practical farming. It is possible to distinguish many relatively small areas of differentiated farming patterns,[1] but a view of the whole can only be obtained by accepting a degree of generalization which might not be considered satisfactory if a smaller country were being analysed.

The division of the state into economic as well as natural regions has occupied a central position in Russian geography since the eighteenth century. After the Revolution, several schematic maps of agricultural and other economic distributions were produced.[2] The need for more detailed study of regional characteristics of crop and livestock production to assist agricultural reorganization was stressed in 1930 and the Lenin All-Union Academy of Agricultural Sciences prepared a map showing 44 regions intended as a basis for promoting specialization.[3] From the changeover towards diversification of production in the mid-1930s, Baranskiy used a map of 18 regions until it was considered to be out of date.[4] Eighteen zones were also used for a map of planned agricultural production prepared in the early 1950s.[5] In 1956 new goals were set for agriculture and groups of scientists undertook a detailed programme for regionalization, including the study of physical regions intended to form a basis for

[1] *Atlas sel'skogo khozyaystva SSSR* (1960), *Atlas razvitiya khozyaystva i kul'turi SSSR* (1967), *Atlas SSSR* (1969), etc.
[2] Nikishov (1960), 24–25.
[3] Jackson (1961), 659–666.
[4] Jackson (1961), 666–669.
[5] Nemchinov (1955); Jackson (1961), 674–676.

agricultural planning and land use.[1] A map of farming zones with special emphasis on management requirements distinguished 47 areas,[2] while 32 regions were recognized in a study of agricultural specialization.[3]

There is thus a wide range in Soviet studies of agricultural regionalization. Though all appear to lack an adequate statistical basis, adherence to one of the Soviet schemes of regionalization in any description of agricultural regions in the present work appeared desirable for a number of reasons. In the first place, they have been derived after prolonged study of the available data; secondly, these figures are obviously much more extensive than those currently available in the west; thirdly, the major regions described represent generalizations from a network of regions in greater detail. Some characteristics of these more detailed regions can be seen in atlas maps, which depict crop and livestock combinations. Regions in greater detail are also illustrated by Rakitnikov.[4]

Rakitnikov, as well as paying particular attention to the methodological problems of agricultural regionalization,[5] has provided maps and descriptions of the major regions of the country's agriculture for current Soviet studies of economic geography.[6] For this reason, Rakitnikov's scheme has been used as a basis for this chapter. In the map (Figure 20) and accompanying regional description minor alterations have been made to Rakitnikov's scheme. Thus, Soviet textbook maps show one mountainous region in the Far East which is not dealt with in their descriptions and this has here been amalgamated with the neighbouring valleys which provide the focus for the whole area. The principal areas around all the major cities are separately distinguished as one 'region' in the Soviet maps because of the development of dairying, market gardening and other specializations to serve the urban areas. On the scale of the generalized treatment, this seems

[1] Gvozdetskiy (1960).
[2] Sotnikov (1960).
[3] Nikishov (1960).
[4] Rakitnikov (1970).
[5] For example, Rakitnikov and Kryuchkov (1966), Rakitnikov (1970).
[6] For example, in Nikitin, Prozorov and Tutykhin (1966) and Saushkin, Nikolskiy and Korovitsyn (eds.) (1967).

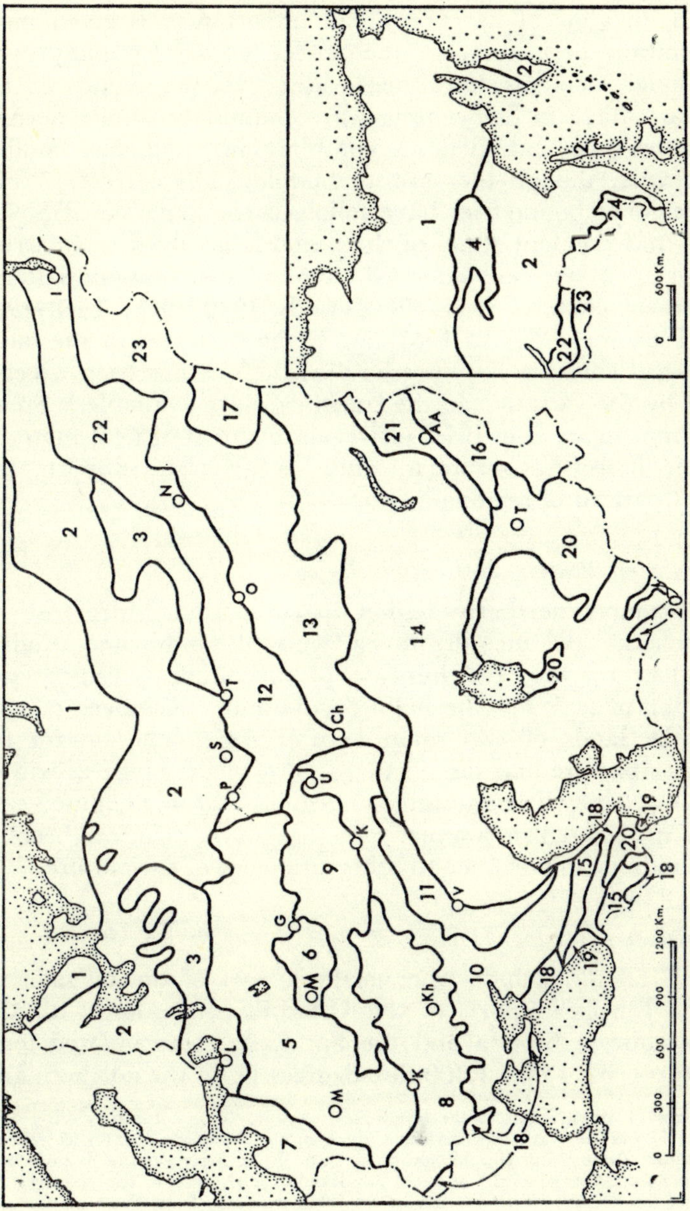

20. Agricultural regions (after Rakitnikov and others)

unnecessary except in the case of the area around Moscow, which, in view of its exceptional importance, is given more recognition here than in the Soviet texts. In many cases locational elements have been added to the names of the regions which, in Soviet usage, are commonly wholly derived from the crop and livestock combinations, and may confuse the reader, though they are methodologically correct. While the regional boundaries have followed the generalized Soviet maps, independent study of the patterns involved was carried out using about one hundred maps of distribution patterns and their changes over the periods 1940–1958, 1958–1965 and 1965 (or 1960) to 1967–69. Some variation in the dates and areal coverage for specific distributions has been necessitated by the variation in the statistical data available.[1] Space precludes more than brief reference to this statistical material which, however, amplifies, and, in general, supports the Soviet pattern of regionalization.

(1) *Reindeer Rearing & Hunting Region*

This most northerly region extends from Murmansk to Kamchatka and includes tayga, wooded tundra and tundra. Reindeer rearing, the hunting of fur animals, fishing and catching of seals are the main occupations. Reindeer rearing uses the lands of two main natural zones—the tundra for summer pasture and the northern parts of the tayga as winter pasture, while in mountainous regions, the high, unforested lands are grazed in summer. To the south, reindeer become of subsidiary importance to other branches of agriculture.

(2) *Reindeer Rearing, Hunting & Agricultural Region*

This region includes the northern part of the tayga zone in the European parts of the U.S.S.R., and almost all the tayga zone in Siberia and the Far East. Accumulated temperatures are 1,000–1,400 day-degrees C. in the northern and

[1] Maps of distributions in the R.S.F.S.R. were made by computer, using data for the 71 oblasts and other divisions given in the *Narodnoye khozyaystvo RSFSR* handbooks, these being supplemented by data for major economic regions and republics from the general statistical handbooks for the rest of the country. All statistics for areas and production in this chapter come from these handbooks. Page references are not given for reasons of space but most can be traced through the footnotes to Chapters 7, 8 and 9.

1,600–1,800 in the southern areas, with frost-free periods of from 70 to 120 days.

Some cultivation and rearing of cattle, pigs and sheep occurs in islands of improvement in forest and marsh. Heavy dressings of fertilizers are needed. Agricultural work is commonly combined with forestry, fishing and hunting.

The most developed parts of these two northern regions are in the European north-west areas. Most of the agricultural production of Murmansk oblast is derived from region 2. Throughout the 1960s Murmansk had twice as many cattle as in 1941, but the total was only about 15,000, including 10,000 cows. At the beginning of 1969 pigs numbered 23,000 compared with 6,000 in 1941, but sheep had declined from 15,000 to 6,000 in the same period. The oblast also claimed about 80,000 reindeer (70,000 in 1941). The Komi A.S.S.R., also in region 2, had 144,000 reindeer in both 1941 and 1969 but it showed a reduction in cattle and sheep over the period as well as in grain and other crops, illustrating the problems of development in these northern areas and the need for stimulation of nearby markets if increased production is to be achieved and sustained.

(3) *Northern Dairying Region*

Within the tayga zone, a more developed form of agriculture is found in some areas. This is true of much of Leningrad oblast, the southern parts of Karelian A.S.S.R., parts of Arkhangelsk oblast, including the Northern Dvina and other valleys and most of the south of the oblast, and adjoining parts of Komi A.S.S.R.

Specialization is generally in dairy-beef cattle raising, based on the natural fodder of areas of water meadows and low-lying pastures. Cultivation has subsidiary significance, but fast-growing and frost-resistant varieties of grain (mainly barley and rye) are grown, and a larger share of the arable area is occupied by potatoes and fodder crops. Extensive areas of marsh exist.

Leningrad oblast is the most highly developed part of this region, the influence of the city market having led to increases in milk output in the oblast of 62 per cent. above the 1940 figure by 1958 and a further 17 per cent. by 1965, achieved almost entirely by improvements in yield per cow. Meat production in the oblast increased yet more rapidly, by 84 per cent. and 47 per cent. in the two periods, mainly as a result of increased numbers of pigs. In neighbouring Karelia, also benefiting from the growth of Leningrad and Murmansk, milk production in 1968 was over three times greater than in 1940, and meat production was double. More potatoes and vegetables were grown but the total sown area was 40 per cent. less than in 1940, grain cultivation having been practically abandoned.

Similar but less developed dairy-beef cattle husbandry is found in the valleys of the Ob, Tobol and Tara east of the Ural mountains (most of Tomsk, north part of Omsk and southern Tyumen oblasts), where extensive water meadows occur in these better drained areas of the west Siberian lowlands. Milk output has been increased rapidly but meat production has, over the whole post-war period, risen more slowly, the figures for Tomsk oblast being 1940–58, milk 76 per cent., meat 32 per cent., and 1958–65, 13 per cent. and 15 per cent. respectively.

(4) *Livestock Rearing & Cultivation Region of Yakutia*

In eastern Siberia more continental conditions prevail and soils are less leached. In the central Yakutia lowlands livestock are reared on natural pasture and hay found in forest clearings, with increasing supplementation by fodder crops. Although grain crops can be grown at these unusually high latitudes, because of the winter drought only spring crops are practicable. Barley, spring rye and spring wheat are predominant, but the sown area of Yakutsk oblast decreased by 25 per cent. from 1940 to 1958 and by 18 per cent. from 1958 to 1965. The fodder available and the extensive nature of the husbandry result in a bias towards beef cattle, with dairying subsidiary, and cattle and other livestock numbers have increased substantially since the war.

(5) *Central Dairying & Arable Region*

This region includes the southern part of the tayga and the greater part of the zone of mixed forests west of the Ural mountains—comprising Belorussia and the oblasts of Pskov,

Novgorod, Kalinin, Smolensk, Vologda, Kostroma, Kirov, the Udmurt A.S.S.R., and parts of neighbouring divisions. The sum of daily average temperatures of the growing period increases from 1,600–1,800 day-degrees C. at the northern boundaries up to 2,000–2,200 day-degrees C. at the southern boundaries. Thus, the cultivation of grains is here not seriously restricted by the amount of warmth and the number of plants that can be raised is much greater than in the previously described regions.

The soils—primarily turf-podzols and marsh soils—require heavy dressings of lime and fertilizers and there is still a high proportion of forest and marsh land as yet underdeveloped. Over the greater part of the region arable land varies between 20 and 40 per cent. of the agricultural area. Past reclamation has reduced natural pastures which are now confined to the low-lying areas with water table near the ground level and to other infertile soils.

Most of the sown acreage is occupied by potatoes, grains (with rye most important but including maize and other crops for silage), pulses and root fodder crops. Root crops are used as a fodder base which also restores fertility to the soils, while liming and fertilizing, including the use of peat, are particularly important in this region of surplus moisture and leaching. With their use and with drainage, sown pastures can be made highly productive. A typical rotation pattern for 7–8 fields gives two fields of perennial grasses, 3–4 fields of grain and pulse crops and one or two fields of intertilled crops.

A good supply of pasture and fodder facilitates dairying with low costs and also makes for a wider variety of livestock than in the more northerly regions. This region has been stimulated by large urban and industrial centres, including Moscow, and relatively good communications, which have promoted intensification of agriculture, particularly in dairy, pig, poultry and vegetable production. Pig rearing is based on potatoes, grain and the waste products from milk processing. Many large sovkhozes have been created in this region, especially near Moscow, to meet the demand for meat, milk, vegetables and fruit.

Further from industrial centres, farming combinations include flax and potatoes with livestock rearing. This is the most important region for flax—Kalinin, Smolensk, Pskov, Vologda and Novgorod oblasts having the largest areas sown with this crop in 1967 in the R.S.F.S.R.

(6) *The Moscow-Gorkiy Market Gardening & Dairying Region*

The influences exerted by Moscow on the preceding region are seen to much greater effect in this area, immediately surrounding the capital and extending westwards to include Gorkiy. Despite the industrialization, much land still remains under forest and agricultural land amounts to about half the total area, about 60 per cent. of this being arable. Little change has occurred in the cropped area compared with the pre-war period but rye is grown much less than formerly, its place being largely taken by wheat e.g., Moscow oblast; 1940, rye 185,300 hectares, wheat 72,100 hectares; 1965, rye 81,500, wheat 163,800. Fodder crops have been roughly doubled in area and much more land than formerly is sown to grass to maintain about 75 per cent. more cows than in 1940. From 1965 to 1968 cow numbers changed little but milk output continued to rise as yields improved in the specialized dairy farms serving the cities, and the total output in Moscow, Vladimir and Ivanov oblasts totalled in 1968 three times the 1940 figure. Total meat production has also risen to rather more than twice the 1940 amount despite falls in the numbers of pigs and sheep as stress has been put on cattle. Egg production in Moscow oblast in 1968 was nearly five times that of 1940.

The light soil of the Meshcherskiy area south of Moscow has been noted for specialization in potatoes as well as dairying and this remains an important enterprise, but the area devoted to potatoes in Moscow oblast and most neighbouring areas has been reduced, while flax has been much more sharply cut back as food for the urban population has received priority. In 1968 Moscow oblast had only 27,100 hectares in vegetable production compared with 38,200 in 1960 and 41,900 in 1940 but the 1968 output was 781,000 tons compared with 662,800 in 1960 and 362,200 tons in 1940, indicating the growing intensification of production. Light grey soils have proved suitable for a wide range of fruit trees.[1]

[1] Kolesnikov (1967).

(7) *Baltic Dairying & Pig Rearing Region*

This region includes the greater part of the Latvian, Estonian and Lithuanian republics, the western part of Belorussia and the Kaliningrad oblast of the R.S.F.S.R.

Climatically, this most western region of the forest zone enjoys relatively moderate temperatures. Lithuania and the coastal belt around the Gulf of Riga have a frost-free period of more than 150 days. The sum of average daily temperatures ranges from 1,600 to 2,200 day-degrees C. The July mean temperature is 18–19°C, and annual rainfall exceeds 600 mm. Soils are turf podzols with a significant area of light sandy loams and sandy soils.

The chief crops are potatoes, fodder crops and bread grains. Potatoes are produced for industrial purposes and as a food crop, but above all for fodder. Natural pastures are less important, arable and permanent hay meadows more important than in region 5. Marshlands have largely disappeared under cultivation. There has been a marked development of dairying, especially in Lithuania where in 1968 cattle numbers were 60 per cent. higher and milk production 85 per cent. higher than in 1940–41. Pigs are important and in 1968 were about 60 per cent. higher than in 1941 but below the peak figures of the early 'sixties. Meat output from all classes of livestock and egg production were at peak levels in 1968.

(8) *South-central Sugar Beet, Grain & Animal Husbandry Region*

This region occupies the western parts of the wooded steppe zone and extends into the southern parts of the mixed forest and northern parts of the steppe zones, comprising the Central Chernozem economic region, and parts of the south-west Ukraine and Donets-Dnepr economic regions. Pasture and hay meadows are found in moist lowlands, in river valleys and among the mixed forests, but further south less of the land is occupied by forests, waste-lands and commons, and more is cultivated. Of the agricultural area, 75 to 80 per cent. is under the plough.

The whole wooded steppe zone is noted for the evenness of the water balance, though the eastern regions are more continental. Precipitation varies from 440–450 mm. in the south-east of this area to 550–575 mm. in the north-west, but severe droughts occur once or twice in four years and so irrigation is valuable. The soils, though rich, are easily washed or blown away and soil erosion is a hazard to be guarded against continually and shelterbelts are planted to provide protection.

The sum of average daily temperatures in the vegetative period amounts to about 2,400–2,500 day-degrees C. in the north and 2,800–3,000 day-degrees C. near the southern borders of the region, while average frost-free periods are 150 days, rising to more than 160 over much of the region. The dominant soil types are grey forest soils, leached chernozems and typical chernozems in the wooded steppe and chernozems in the steppe.

Because little natural pasture is available, the scale and character of livestock rearing is based on cropping. The majority of farms combine grain growing and livestock rearing.

The development of the virgin lands enabled the area under grain in this and neighbouring regions to be reduced and in 1967 in this region grain occupied about 50 per cent. of the arable area compared with 60–75 per cent. in 1940 and 1950. Winter wheat was allocated more land (in the Central Chernozem region, three times as much in 1967 as in 1950) but spring wheat and rye were cut back sharply. State purchases of grain, especially wheat, rose sharply throughout the region despite the smaller areas sown.

Sugar beet is the leading industrial crop, occupying about 6–10 per cent. of the total sown area and 25 per cent of some farms.[1] This is 3–4 times as much as in 1940 in the Central Chernozem economic region but in the Ukraine there has been only a small increase and this is one way in which the differences between the two parts of the agricultural region have been reduced. Sugar beet requires intensive cultivation through deep ploughing, frequent loosening of the soil between rows and heavy fertilizing, and following crops gain from the weed-free fields. Maize, for which good conditions prevail, has long been important in the western part of the region and has become more widely grown in the east.

[1] Saushkin, Nikolskiy and Korovitsyn (eds.), 1967, 340, but see also p. 196, quoting Hultquist and other sources.

Sugar-beet regions have relatively high intensity livestock rearing. The mass of waste products of the beet processing, maize and production of concentrates together ensure the fodder supply, and the region has become important for pre-slaughter fattening. Whereas cattle from the pastoral fattening regions reach the slaughter houses predominantly in autumn, here the main fattening period is that of the beet processing, i.e., winter. Dairying is well developed with the leaf and waste (*zhom*) of the sugar beet much valued for fodder, while concentrates and good fodder also help pig rearing. All told, the region is carrying more than twice as many cattle and pigs and about the same number of sheep as in 1940.

In the southern, drier parts, good crops of winter wheat require clean fallow in the rotation, and sunflower is an important crop, Voronezh oblast having the third highest output of the crop in the R.S.F.S.R. after Krasnodar kray and Rostov oblast in 1967. An 8–9 field rotation is employed, with 55–60 per cent. of grain and bean crops.

(9) *East-central Region of Arable & Animal Husbandry*

This region is also within the wooded steppe and mixed forest zones but further north and east than the preceding region, extending from the mid-Russian heights to the Ural mountains, taking in parts of the Central, Volga-Vyatka and Volga regions. The climate here is much drier. Accumulated temperatures for the vegetative period are markedly less—about 2,100 day-degrees C. at the northern boundary and 2,500 day-degrees C. at the southern. The average frost-free period is shorter—from 140 to 150 days over most of the area. Soils are vulnerable to erosion and attention to this matter has been much increased in recent years.

Crop rotation is based on bread grains, maize and potatoes. Grain crops occupy 50–60 per cent. of the sown area. Formerly the main crops were winter rye, oats and 'groats crops'—millet and buckwheat. In recent years the emphasis has turned to wheat, with increases of 20–60 per cent. in the wheat area of most parts between 1958 and 1965. Characteristic crops of parts of this area are potatoes and hemp. Potatoes are grown as a commercial crop mainly in the north of the region in the zone of grey forest soils. Hemp, on the other hand, is grown for preference on loamy soils in competition with other intensive crops, such as makhorka and vegetables, in the warmer areas. Sugar beet is widely grown and sunflowers occupy a comparable area on suitable land in the east. Fruits, especially apples, also provide a distinctive element.

In 1965, fodder crops occupied 2–3 times the 1940 area in most parts, 5–8 times in some. Livestock rearing is based largely on these, the most important branch being cattle, but with increasing attention to pig breeding and poultry keeping, which need less pasture. Numbers of livestock in some parts of the region, such as the Chuvash and Bashkir A.S.S.Rs, reached record numbers in 1969, though in some oblasts there was a slight decline from peaks reached in 1967.

(10) *Central Grain & Animal Husbandry Region*

This region extends from the Bashkir A.S.S.R. to the Black Sea and north Caucasus and includes most of the European steppes. Accumulated temperatures in the vegetative period amount to about 2,500 day-degrees C. in the northern parts and 3,500 on the Caucasian foreland.

The advantages of greater warmth are for many crops offset by the adverse water balance, so that harvests tend to be less than in the wooded steppes. Irrigation is necessary for good yields of the more moisture-sensitive crops. Mechanization and technical equipment of a high order are prerequisites to development of this region, in which conservation of moisture is critical.

There have been great advances in overcoming drought; protective forest plantations, snow control, construction of ponds and reservoirs, irrigation from local streams and also from the great rivers—Don, Kuban and Dnepr. On the irrigated lands there has been development of sheep rearing, orchards, vineyards and, in places, rice growing. The development of virgin lands resulted in extension of the arable area, and in 1968 this amounted to 60–80 per cent. of the total land classed as agricultural. By 1958 the sown areas in individual oblasts were up to 50 per cent. higher than in 1940 and between 1958 and 1966 further increases up to another 12 per cent. were recorded. These increased areas of cultivation have generally been maintained, though in some parts there have been reductions, e.g. in Stavropol kray from 4,567,000 hectares in 1965 to 4,232,000 in 1968 consequent upon restoration of fallowing.

Grains occupy generally 50–70 per cent. of the sown area, rising to 76 per cent. in Orenburg oblast, with wheat occupying similar proportions of the total grain. In the less dry, western parts of the region maize sowings have been much extended to support a swing towards more intensive livestock husbandry, and wheat is mainly sown in winter. Eastward, the climate becomes harsher and spring-sown varieties progressively replace winter wheat. Sunflowers are the most important industrial crop, with sugar beet also favoured in some areas, notably Krasnodar kray. Also in the less dry areas and on land with a high water table southern hemp is grown. Melons, grapes and other fruits are produced. Fodder grain, silage and other fodder crops (notably maize and fodder melons) are grown in great quantities. These permit relatively low cost dairy-beef cattle, pig and poultry rearing despite grass being restricted by the climatic conditions. Stavropol kray retains its importance in specialized sheep breeding. Here and elsewhere in the eastern parts, unimproved steppe pastures remain over extensive areas and are used for stock rearing.

(11) *South-central Grain & Livestock Rearing Region*

This region, comprising parts of Stavropol kray and Saratov, Volgograd and Rostov oblasts, is semi-arid and has dark chestnut soils. The agricultural emphasis is on grain crops and livestock rearing, but vegetable crops, melons and mustard are important. The lower Volga valley provides a zone of fertile and intensively cropped soils separating semi-desert areas to east and west. Elsewhere, solonets and other soils unsuited for cultivation explain a proportion of pasture rather higher than in region 10. Cattle, mainly beef, and fine-woolled sheep utilize these lands. Large numbers of sheep from the neighbouring parts of the steppe zone are moved for winter grazing to the arid steppe pastures of the pre-Caspian lowlands.

(12) *Eastern Cattle & Grain Region*

This region stretches from the Ural mountain valleys of Perm, Sverdlovsk and Chelyabinsk oblasts across the west Siberian plains to Novosibirsk. Accumulated temperatures for the vegetative period are about 1,800 day-degrees C. in the north and 2,100 day-degrees C. in the south, with a frost-free period of 110–120 days. The European part is the more developed, with a cattle and grain type of agriculture in the eastern part of the forest zone, within the Ural economic region. There is less forest land east of the Ural mountains and the wooded steppe is the main type of terrain used for dairying and grain husbandry.

The better drained areas are in arable cultivation and the abundance of land led to major developments under the virgin land schemes so that this became an important grain producing region with spring wheat as the chief crop. Large areas of marshes and saline soils continue to hinder development but some are utilized for dairy and beef cattle.

(13) *Eastern Grain & Livestock Rearing Region*

From the southern part of Orenburg oblast and the neighbouring north-east part of Kazakhstan the steppes sweep eastward to Novosibirsk and Altay kray. Accumulated temperatures for the vegetative period range generally from 2,100 to 2,600 day-degrees C., and to 3,000 in the extreme west of the region. The frost-free period in the east is less than 130 days. Annual precipitation ranges from about 250 mm. in the south to about 350 mm. in the north, with a summer maximum. Soils range from southern chernozems to dark chestnut with considerable areas of solonets types unsuited to cultivation.

Widespread development here began only at the end of the nineteenth century, but between 1954 and 1960 it was the scene of the main development of virgin lands. Large areas were developed in the chernozem steppe, particularly by new state farms organized on land previously in pastoral use. The biggest continuous areas of development were, however, in the dry steppe characterized by the dark chestnut soils, and it was here in Tselinny kray (virgin land territory) that the greatest number of state farms was formed to carry out the transition from pastoral to arable husbandry.[1]

The region has developed the most specialized grain farming in the country, surpassing all other regions in the area cultivated

[1] For the performance of the decade of development, including initial successes and subsequent setbacks, see Zoerb (1964).

per man. Although the climatic conditions result in sharp fluctuations in yields, the 1966–68 average total yield for all Kazakhstan was over 16 million tons compared with 1·6 million tons in 1940, while the Ural and west Siberian economic regions averaged about 12 million tons compared with 3 million in 1940. Most of these increased outputs for the administrative areas come from this agricultural region. Part of this region, adjacent to the Altay, enjoys significantly higher and more regular yields. Accumulated temperatures of 2,200–2,300 day-degrees C. combined with higher rainfall permit a measure of concentration on sugar beet. Livestock rearing has been intensified as improved fodder supplies have become available.

(14) *Pastoral Husbandry Region of the Desert & Semi-desert*

With the transition to semi-desert conditions in Kazakhstan, Astrakhan oblast and the Kalmyk A.S.S.R., natural pasture becomes the main form of land use in non-irrigated areas. In the south of the region snow cover is light and of short duration so that pasture provides maintenance for livestock throughout the greater part of the winter, whereas in the semi-desert conditions between the Aral Sea and Lake Balkhash, snow cover lasts 80–120 days and stall feeding is necessary.

Large areas with sandy soils offer comparatively good pasture in spring when the ephemerals are growing and retention of water in the sands makes possible continued grazing in the summer, but feed falls off as the water is used or evaporated. Conditions improve again in autumn when absinthe and other plants yield large quantities of fodder. Sheep rearing is generally dominant but cattle are important in such areas. Many are driven in summer to the heights of the Mugodzhar, Kazakh foothills, Tarbagatay and Tyan-Shan ranges to graze mountain steppe, meadow-steppe and meadow pastures.

Contemporary husbandry is based on the state and collective farms and is not nomadic, in contrast to the form of livestock rearing which previously existed here. Hay is prepared in grazing areas and fodder concentrates are transported from arable regions. Water supply to pasture and irrigation of hayfields is increasingly ensured by modern techniques, including utilization of artesian water. The production centres of the farms and permanent settlements are usually located near the wintering places.

In the southern desert zone pastures suffer more from water shortages and fodder is consequently very poor in summer. Here, breeding of Karakul sheep has become important, but sheep farming is subsidiary to camel rearing.

(15) *Mountain Livestock Rearing Region of the Caucasus*

In the mountain regions, where cultivation is hindered by relief and shortness of the growing season associated with altitude, livestock rearing is dominant. The transition takes place at heights varying from 700 to 1,500 m. Pasture land is less prevalent in the western, more forested and moister parts of the Caucasus than in the eastern, drier parts.

Cattle are brought for summer grazing from the foothills and plains (cotton, grain, orchard-vineyard and tobacco regions), and, in addition, there are livestock rearing collectives based on the mountain lands themselves. Both forms of utilization are necessary since the grazing capacity of the mountain lands is much higher in the 3–4 summer months than in the rest of the year. In fact there is two-way interchange, livestock from the mountain farms being taken to the plains in winter. This seasonal exchange extends over a distance of 300–500 km. Sheep are involved in the furthest movement, being pastured in summer on the high, stony and less well grassed areas, while cattle are kept on the less steep and better lands.

(16) *Mountain Livestock Rearing Region of Central Asia*

This region includes the high mountain areas of the Kirgiz and Tadzhik republics, on both ridges and intermontane basins with altitudes above 1,500 m. The pastures are of steppe rather than meadow type, and the absence of winter precipitation in the intermontane basins limits fodder, Because of this, in contrast with the Caucasus, the livestock rearing is much more concerned with wool and meat than dairying.

Wool output has been greatly increased as a result of the crossing of the formerly widespread Kurdish sheep with the fine-woolled Tyan–Shan breed. On the Pamirs and in the highest intermontane valleys of the Tyan–Shan, yaks are found in addition to sheep and horses.

(17) *Mountain Livestock Rearing Region of the Altay*

In the inner parts of the Altay, shut off from the west and north by mountain ranges, the climate is dry, and the dominant vegetation is grassland of meadow-steppe, steppe and, in parts, even semi-desert type. Winter precipitation, in particular, falls off sharply.

In the central, eastern and south-eastern valleys of the Altay, pasture is overwhelmingly the main form of agricultural land use. Beef cattle and sheep, together with yaks are reared. In the north and north-west, cattle take precedence and, while beef is the main objective, cheese and butter factories have been established.

(18) *Orchard, Vineyard & Tobacco Growing Region*

The conditions which encourage these crops include high accumulated average temperatures (3,000–4,000 day-degrees C.) a long frost-free period (190–250 days) and a mild winter (average January temperature above zero). These conditions are found in Transcaucasia and on the coasts of Dagestan, and from the Krasnodar coastal lands to the Crimea with inland areas (rather cooler) in Moldavia and Transcarpathia. Where the relief hinders cultivation of field crops, it helps towards freedom from frost, good soil drainage and a choice of differently exposed slopes, all valuable for fruit cultivation. Fruit is also dominant on the flats among the foothills where conditions are suitable for artificial irrigation, using the many mountain streams.

Apples, pears, cherries, peaches, apricots, figs, walnuts and grapes of both wine and table varieties are grown. Vineyards are a specialization of the areas with drier summers. Tobacco is important in the wetter areas with leached soils in Transcarpathia, the Moldavian republic, the Crimea and western parts of the Caucasus.

In these areas fodder for cattle is limited. However, on hill slopes and in the dry areas of the eastern Trancaucasus and Dagestan, pastures are available and cattle and sheep rearing has developed with seasonal movement as necessary. Pigs are also kept, in some cases on forest pastures. Silkworms are important in some areas, the cultivation of the mulberry and feeding of silkworm caterpillars integrating well with the local intensive agriculture.

(19) *Region of Sub-tropical Perennial Crops*

This region includes the hilly Caucasian mountain forelands of the Adzhar and Abkhaz Autonomous Republics fronting the Black Sea, and the cultivated parts of the Kolkhid and Kura lowlands. It is the principal region in the U.S.S.R. classed as sub-tropical. Accumulated temperatures in the vegetative period exceed 4,000 day-degrees C., there are 240–250 frost-free days and average January temperatures range from 3 to 8 degrees C. Annual precipitation totals about 1,300 mm.

The cultivated lands occupy little of the total surface area, which includes both high mountains and marshy lowlands. In 1968 tea occupied 63,100 hectares in Georgia and 7,300 hectares in Azerbaydzhan. Plantations extend to 700–800 m. on the hillsides and also over a considerable area on the plain. Widespread cultivation of tea is facilitated by its frost-hardiness and leached soils are essential.

Citrus fruits are more tolerant of soil conditions but restricted climatically, particularly by winter minimum temperatures. Plantations of lemons, oranges and mandarins are located mainly near the sea and on steep slopes where the risk of frost is least, frequently on terraces. Another important sub-tropical crop is tung oil. Tung trees are less frost-hardy than tea but more resistant than citrus fruits, and require less labour input.

Among annual plants, tobacco is important, its distribution depending on soil characteristics. Essential oils (chiefly geranium) have a limited distribution on the plains. Vegetables grown in the cooler seasons can be marketed earlier than those from most other regions.

The chief field crop is maize. There is little land in pasture but some collectives drive cattle to the sub-alpine and alpine pastures in the summer. Far more important in value, however, than ordinary livestock rearing, is the breeding of silkworms.

(20) *Regions of Cotton Growing & other Irrigated Crops*

These are the regions of Central Asia and Transcaucasia with accumulated temperatures of over 3,800 day-degrees C., frost-free periods of 180–230 days and rainless summers with a very high number of sunny days (up to 26 in July and 28 in August). The mountains provide water for artificial irrigation.

Within the zone climatically suitable for cotton there are regions both of strong specialization in it (including Fergana, the Hungry steppe and the main part of the Zeravshan valley) and of considerable development of other branches of agriculture. Where irrigation systems permit an abundance of water in the summer, cotton is generally given precedence, with lucerne grown in rotation with it. Rice is the dominant crop in the middle and lower reaches of the Syr Darya. Maize and jute are also important field crops and vineyards, orchards and sericulture are subsidiary enterprises. Both mineral fertilizers and manure are used in large quantities, while the irrigation waters themselves deposit silts which contain nutrients.

The use of non-irrigated lands is in many cases integrated with the irrigated areas. In the foothills, many cotton collectives grow crops capable of using the abundant rain in early spring and finishing their growth by the onset of the dry season. Such crops (*bogarnyye posevy*) comprise wheat (mainly winter varieties) and barley (spring and winter varieties) but also include flax, mustard and sesame.

Cropping of this type has great value to the farms for the supply of coarse and concentrated fodder. Livestock rearing in the cotton region is, in general, divided into two isolated types, that of the oases, based largely on the surrounding pastures, and that of the desert plains of Central Asia where flocks comprise predominantly Karakul sheep.

This region has great potential for development because of its reserves of land and sources of water. The great rivers such as the Kura, Syr Darya and Amu Darya have large reserves of water not yet exploited for irrigation. The greatest opportunities are in Central Asia in the middle reaches of the Syr Darya and north of the Amu Darya in the Karshinskaya steppe and in the lower Zeravshan valley, while in Turkestan the Amu Darya-Kara Kum canal offers great possibilities. In Transcaucasia the Mingechaursk hydro-electric scheme in Azerbaydzhan provides the basis for utilization of the waters of the Kura river.

(21) *Central Asian Regions of Intensive Crops & Subsidiary Livestock Rearing*

North-east of the cotton region, in Kazakhstan and Kirgizia, an area important for crops slightly less warmth-demanding than cotton extends from the foothills of the Tyan-Shan down to about 400 m. on to the plain.

On the edge of the plains the accumulated temperatures for the growing season amount to about 3,500 day-degrees C. and precipitation varies from 120 to 250 mm. per annum.

In irrigated areas beside the Talas, Chu, Ili, Karatal and other river systems, crops include sugar beet, rice and other grains and fodder crops to facilitate livestock rearing. Southern hemp and kenaf are locally important, notably in the Chu valley. Favourable climatic and soil conditions have encouraged the development of horticulture on the foothills, especially near Alma Ata. Grain-livestock combinations are found on higher land where sufficient precipitation coincides with gently sloping land.

In the dry non-irrigated areas land is largely semi-desert pasture useful for winter, spring and autumn grazing. Fodder, including hay, is obtained from the foothills above 500 m., while higher yet is the zone of mountain pastures. These are used chiefly in summer but some are available for grazing also in winter, especially those in valleys and ravines sheltered from the north and west by mountain ranges and therefore little affected by snow. Settlements are commonly located in irrigated areas.

In all parts of these regions livestock rearing is based on the seasonal availability of different pastures, alpine and sub-alpine in summer, semi-desert in winter, though these involve movement of stock over considerable distances.

(22) *Grain & Livestock Region of Eastern Siberia*

A grain and livestock region extends eastward discontinuously from Kemerov oblast through the southern parts of Krasnoyarsk kray, Irkutsk oblast, the Buryat A.S.S.R. and Chita oblast. Agriculture is found in the warmer intermontane basins, separated by tayga-covered heights.

These areas show fairly high degrees of continentality with

most precipitation in summer and only light snow in winter. Accumulated temperatures for the growing season range from 1,500 to 2,500 day-degrees C. with the water balance also very variable. Over considerable areas forest is the natural vegetation, but in valleys and basins sheltered by mountains and accordingly dry, there are many areas of wooded steppe, herbaceous steppe, and even dry steppe in Chita oblast and the Buryat A.S.S.R.

The character of the agricultural economy strongly reflects the great distance of these regions from the economically advanced and densely populated regions of the U.S.S.R. Their economy is founded on the supply of the most easterly regions with products for which there is a continuing demand. The emphasis is on grain, meat and dairy products and wool with relatively little regional differentiation.

Virtually all the agricultural land of Krasnoyarsk kray and Irkutsk oblast is in this region, and these areas illustrate its development. The sown area was 4·9 million hectares in 1968 compared with 2·8 million in 1940, and the area devoted to grain 3·2 million compared with 2·4 million. More than twice as much grain was harvested in each of the years 1966–68 as in 1940. Cattle numbers in 1968–69 were 66 per cent. higher than in 1941, pigs 85 per cent. and sheep 31 per cent. higher. These, however, are not particularly impressive improvements for a period of three decades. Improved milk yields made possible a 1968 production nearly three times that of 1940 while egg production was four times as high.

(23) *Livestock region of the Far East*

In the Sayan and neighbouring mountainous areas of Tuva A.S.S.R., the Buryat A.S.S.R. and the southern part of Chita oblast, agricultural land is limited to the larger valleys, and cattle and sheep provide the basis for the local economy. As snowfall is light, grazing can be prolonged late into autumn and even into winter, but hay and fodder crops are vital. Some fields are irrigated. Tuva illustrates the region. Only 12·8 per cent. of its agricultural area was used as arable land in 1968, and only 2·3 per cent. for hay. Cattle numbers, at 194,000 were 40 per cent. higher than in 1961 but sheep and goats had increased in the same period only from 873,000 to 1,068,000.

Meat, milk and wool outputs were all, however, reported about 60 per cent. higher than in 1961.

(24) *Grain, Rice & Livestock Region of the Far East*

This region occupies the lowlands of the Amur and Ussuri basins—the southern parts of Amur oblast, Khabarovsk kray (including the Jewish A.O.) and the lowlands of Maritime kray. Accumulated temperatures reach 2,500 day-degrees C. in the growing season and there is ample rainfall (400–600 mm.) which falls mainly in summer, with a comparatively dry spring. These conditions suit warmth-loving and late-flowering crops, such as soya beans, maize and sorghum, as well as wheat and fodder crops. Good conditions for irrigation facilitate widespread rice growing.

As almost all of the agricultural land of the administrative areas named above is included in this region, statistics for them illustrate its development. Of a total agricultural area of 4·4 million hectares, 2·6 million were arable in 1968, the balance about equally divided between hay and grazing. The 1968 figures (with 1940 in brackets) were: sown area 2·5 million (1 million) hectares, including grain 1 million (730,000) hectares, with grain output 1·5 million (655,000) tons. In the same period livestock products increased much more; meat from 28,400 to 106,500 tons, milk from 200,000 to 832,000 tons, eggs from 53·6 million to 380·6 million.

The rapid industrialization of the Far East and the distance from other regions make it important for it to become largely self-supporting in agricultural products and, clearly, some progress towards this end has been made.

CHAPTER 11

Farm Improvement and Conservation

One advantage claimed for collectivization when it was introduced was that it would facilitate the modernization, and especially the mechanization and electrification, of the countryside. Units too small and too fragmented for economic utilization of machinery would be eliminated and the distribution of machinery and education of farming people in its use would be simplified. Even so, mechanization was an immense problem and there was justification on practical grounds for concentrating the key machinery in Machine-Tractor Stations, apart from facilitating political control over the peasants, who had shown themselves capable of sustained resistance to collectivization.

In 1916, according to current Soviet figures,[1] under 1 per cent. of the energy used in agriculture was obtained mechanically. Tractors were imported, mainly for the southern grainlands. The grain trade had also stimulated the construction of elevators, the first being built in 1888 at Yelets by the local zemstvo. By 1900 there were 62 elevators, but development had not kept pace with the needs even of this branch of agriculture, which was stimulated directly by the export business.[2] Imports of tractors continued after the Revolution, a total of 88,000 being imported between 1921 and 1932. Only 2,700 Russian tractors had been produced by 1928, but 94,300 were made between 1928 and 1932 and production increased slowly to about 100,000 per year after imports ceased.[3]

The first Five Year Plan (1928–33), placed considerable emphasis on production of agricultural machinery. This was essential if food was to be released for human consumption from

[1] *Strana Sovetov za 50 let*, (1967), 152.
[2] Rubinow (1908); Miller (1926), 58.
[3] *Sotsialisticheskoye sel'skoye khozyaystvo SSSR* (1939), 12.

the demands for feeding animals for haulage. In 1928, 32–35 per cent. of all feed was required for draught livestock, a figure reduced to 10 per cent. by 1959. The reduction amounted to about 40 million tons, sufficient to produce about 6 million tons of pork (live-weight) or 35–40 million tons of milk.[1]

The second Five Year Plan (1933–38) made great advances. Existing agricultural machinery plants were expanded and new ones set up. The U.S.S.R. claimed to be the largest producer of tractors in the world, though the total output was exaggerated by quoting available mechanical resources in terms of 15 h.p. units, and production was biased towards heavy, powerful machines. During the Second World War immense destruction of factories and machines made spare parts an increasing problem. Production was increased in the east, for example, by a tractor plant in the Altay region, a combine harvester factory at Krasnoyarsk and other plants in western Siberia, Kazakhstan and the Ural region.

In 1945, total power available for agriculture, mechanical units, horsepower per worker and horsepower per hectare were all below the 1940 levels. By 1950, the situation had been restored and improvement was rapid between 1953 and 1955. Then the delivery of machines to farms slowed down, and deliveries from 1959 to 1961 were not encouraging, though there was more effort to diversify output to match the needs of farms more closely.

The transfer of machinery to the farms with the closure of the MTS stations was both beneficial and disadvantageous for the farms. They got greater control over their sowing, cultivating and harvesting operations, but increased their costs, at least initially. Increased prices of spare parts also made the transition more difficult. In 1962, complaints about the shortage of machines, spare parts and mechanics were numerous.[2]

The poor record in mechanized farming has resulted not only from lack of machinery and spare parts but from failure by the farms to follow maintenance rules, and loss of time caused by machines breaking down in the fields because of workshop schedules being ignored. This has been a common problem. A remedy tried in Altay kray was to issue tractor drivers with fuel

[1] Johnson (1963), 223.
[2] Some of these are quoted by Kabysh (1965), 171–172.

coupons sufficient only until the next maintenance check was due.[1] By 1967 the following position had been reached:

TABLE 25

AVAILABILITY OF SELECTED TYPES OF AGRICULTURAL MACHINERY
(thousands, at end of year)

	1940	*1950*	*1960*	*1967*
Tractors	531	595	1,122	1,739
Combine harvesters, grain	182	211	497	553
of which, self-propelled	—	35	233	548
Beet harvesters	—	0·1	34	58
Cotton harvesters	0·8	4·8	11	39
Silage harvesters	—	—	121	139
Potato harvesters	—	0·1	10	24
Tractor-drawn ploughs	491	519	782	857
Tractor-drawn sowers	306	350	1,003	1,215
Tractor-drawn cultivators	272	317	755	1,027
Lorries	228	283	778	1,054

Source: N.kh. SSSR 1967, 450–460

These numbers are still far below those regarded in 1962[2] as the minimum required for efficient farming. Sowing and the harvesting of grain, other than maize, are almost entirely mechanized, though work is often reported as delayed by insufficient or defective machinery. In most other branches of agricultural work there is still a large gap to be met by new machinery, apart from necessary replacements.

Increasing mechanization is not, however, always welcomed unreservedly. Where there is surplus labour on a collective, improvement in machinery increases the difficulty of providing full employment, and even when farms have the machines some work which could be mechanized may still be performed by hand in order to keep up employment.[3]

The progress in mechanization of various branches of agriculture and the gap remaining are indicated in Table 26.

In livestock husbandry, it was reported that, in 1968, 41 per cent. of cows (55 per cent. on sovkhozes but only 32 per cent. on kolkhozes), were milked by machines, while automatic water

[1] Yu. Borzikov, *Pravda*, 3.5.69, 2; *C.D.S.P.* 21, (18), 31.
[2] *Ekonomicheskaya gazeta*, 1962, No. 35, 13, quoted by Kabysh (1965), 172.
[3] Nimitz (1967), 196.

supply served 66 per cent. of cows on dairy farms and 82 per cent. of pigs on pig farms, though these percentages would not apply on all farms. Mechanized (electrical) shearing was said to apply to 88 per cent. of sheep.[1] These figures show marked improvements on a few years earlier.

TABLE 26

MECHANIZATION OF SELECTED AGRICULTURAL OPERATIONS IN KOLKHOZES AND SOVKHOZES

Per cent.

	1940	*1945*	*1950*	*1960*	*1966*
Grain sowing	61	39	75	100	100
Sugar beet sowing	93	75	92	100	100
Cotton sowing	81	71	92	100	100
Potato planting	4	1	6	58	78
Grain harvesting	47	27	53	92	99
Sugar beet harvesting	—	—	2	54	74
Cotton harvesting	—	—	—	11	29
Potato harvesting	2	0·4	3	34	58
Haymaking	12	8	24	68	81

Source: Strana Sovetov za 50 let, (1967), 158–159.

Complaints about the design of machines are frequent. A new machine is tested by both health specialists and engineers before being put into production. Sometimes successive modifications and tests are required, but machines which have failed certain tests may nevertheless be put into production. In 1965–66, the VPG–4 sugar-beet cultivator was put into production despite adverse reports from several specialists. The Kharkov T–74 was not adequately modified to meet criticism of its cab being stuffy and dusty, and the SSh–75 self-propelled chassis went into production without criticisms having been met.[2] Plants and production ministries in fact do not adhere to the stipulations of the Standard Safety Requirements for Agricultural Machinery and Tools.

Sometimes farms make their own modifications. Thus, the 'fir tree' radial milking machine, introduced about 1954, proved unsatisfactory in general use. In 1966, however, it was reported on a Kuban farm to be milking three hundred cows

[1] *N.kh. SSSR 1968*, 419.
[2] R. Zaitsev, *Komsomol'skaya pravda*, 13.10.66, 2; *C.D.S.P.* 18, (42), 31.

with a quarter of the labour used for the same size of herd on a nearby farm. It had been improved locally and was still used, because, the farm chairman explained, nothing better was available.[1]

Despite the weakness of the present position great progress has been made in mechanizing agriculture. Soviet calculations are that the contribution of horses and oxen to energy harnessed to agriculture fell from 99 per cent. in 1916 to 22 per cent. in 1940, to under 12 per cent. in 1950 and to only 1 per cent. in 1969. Total energy available is estimated to have been 97 per cent. greater in 1969 than in 1960, and nearly five times greater than in 1950. Between 1950 and 1969 the farmed area was greatly increased, but the power available per 100 hectares of sown land nearly trebled.[2]

Specialized machinery for harvesting grain, sunflower, mustard, soya beans, cotton, sugar beet and other crops has received increased attention. Much has been learnt from other countries which have faced a greater problem in shortage of labour, but Soviet agricultural engineers have shown considerable ingenuity. In 1936, the first combine harvester for flax, a crop in which the Soviet Union has a distinctive interest, was produced. Recently special attention has been given to mechanization of fodder crop operations, matching the increased attention now given to diversification of cropping, particularly for livestock production.[3]

ELECTRIFICATION ON FARMS

Only 15 per cent. of collective farms used electricity in 1950, but the percentage had been raised to 49 by 1958, 71 by 1960 and 99·7 in 1969. State farms benefited from earlier electrification: 76 per cent. used electricity in 1950 and 99 per cent. in 1963. Electricity consumed on farms rose from 538 million units in 1940 to 1,538 million in 1950, 9,970 million in 1960 and 33,256 million in 1969.[4]

As with the supply of machinery, there is still much leeway to make up before Soviet farms are as adequately electrified as

[1] G. Radov, *Literaturnaya gazeta*, 18.10.66, 2–3; *C.D.S.P.* 18, (47), 15.
[2] *N.kh. SSSR 1969*, 387.
[3] A summary of mechanization development in the Soviet Union up to 1959 is provided in the *World Atlas of Agriculture*, Vol. 1, 1969, 498–500.
[4] *N.kh. SSSR 1969*, 393.

those in the most advanced agricultural countries. Nor has the present position been reached without much difficulty and local improvisation. In the more remote areas supplies have to be generated on individual farms, and many collectives have developed their own plants, often using water power. Some of these plants have since been replaced by the state grid and some were not in use long enough to justify the original investment.

More seriously, the electrification of agriculture is hindered and made costly by failure of the power authorities to tap the new large power stations. Thus, in Krasnoyarsk kray, one of the richest regions in power resources and development east of the Urals, nearly half the state farms and two-thirds of the collectives in the kray in 1966 were using their own power plants—called 'individual bonfires' by the critics—and only about one per cent. of the power generated by the state schemes was fed into agriculture.[1] These small plants were alleged to employ 10,000 people with cost of electricity eight times that of the state supply system, while over 150,000 kilowatts of installed capacity were unused.

The tapping of high-voltage transmission lines is, of course, always a problem, and farmers in many countries have experienced the frustration of remaining without electricity long after the lines of a regional grid have been built beside them. In the U.S.S.R., however, undue difficulties appear to have arisen from division of authority. In Krasnoyarsk kray it seems that co-operation and integrated planning by the Krasnoyarsk Electricity Grid Construction Trust and the Rural Power Construction Trust, which provides the low-voltage transmission lines, was lacking. Sometimes transmission lines remain idle after construction because local networks have not been completed, at other times the reverse situation occurs.

The electrification of a whole Soviet farm now represents a considerable task. The Bolshevik State Farm in Kurgan oblast, for example, had, by 1966, eleven transformer sub-stations, over 50 kms. of power transmission lines, approximately 3,000 street lights and 400 electric motors.[2] Nevertheless the actual farm processes were still inadequately electrified through shortage of equipment.

[1] A. Morozov, *Komsomol'skaya pravda*, 25.8.66, 2; *C.D.S.P.* 18, (37), 14.
[2] S. Podkorytov, *Pravda*, 8.12.66, 2; *C.D.S.P.* 18, (49), 28–9.

THE FERTILIZER PROBLEM

Shortage of mineral fertilizers has been a major difficulty for the farmers in the Soviet Union, and, as with many Russian agricultural problems, this has a long history. Before the Revolution, manure from livestock provided virtually the only form of fertilizer, though liming and marling of soils was also undertaken locally. The diminution of livestock numbers as a result of war and the resistance to collectivization greatly reduced the amount of manure, and this was reflected in declining yields, especially in the garden crops, to which much of the manure had been applied. Subsequent reduction in the number of horses and working cattle, consequent upon mechanization, has further reduced supplies of this type of fertilizer.

The chemical industry, poorly developed in pre-revolutionary Russia, was not accorded the priority given to coal, steel, machine-building and electrification in the Five Year Plans before the Second World War, nor during the initial years of recovery. In the 'fifties fertilizer supplies doubled but in 1960 were under 11 million tons of all kinds. In 1963, after three years of declining grain yields, Khrushchev revived an earlier plan for rapid development of the chemical industry, emphasizing the production of fertilizers and other farm chemicals. The need was urgent, since the U.S.S.R. was producing only 62 kg. of mineral fertilizers per hectare of arable land compared with the average use of 227 kg. in the U.S.A. and 766 kg. in England per hectare.[1] Khrushchev planned the construction of more than 200 chemical plants and the expansion of some 500 existing factories. In addition to importing machinery from abroad, contracts for constructing whole plants were placed in Great Britain and France. The benefits of this investment are now being realized and by 1969 fertilizer applied per hectare of arable land was 176 kg., nearly three times the 1962 figure.[2] In 1969 total production was about 46 million tons.[3]

Khrushchev also saw the need to improve the distribution of fertilizers. He placed responsibility for allocation to farms with

[1] *Sel'skaya zhizn'*, 18.9.63, quoted by Kabysh (1965), 173.
[2] *N.kh. SSSR 1969*, 357.
[3] *N.kh. SSSR 1969*, 151.

the TPA organization, and both state farms and collectives came under this monopoly, which was subsequently transferred to the Farm Machinery Organization. None of these administrative changes ended complaints and an example of continuing concern was the resolution put to the Party and Council of Ministers in 1967 calling for improvement in supplies of fertilizers and plant protection chemicals, which were reported to be commonly of low quality, poorly packaged and with other faults.[1]

Mineral fertilizers produced in the Soviet Union have hitherto contained only about 20 per cent. of effective nutrients but the expansion of the industry should enable quality as well as quantity to be improved. The distribution while supplies have been short has favoured industrial crops, especially cotton, sugar beet, flax and tea, so with more available there should be a much increased share for grain and livestock production.

There has also been a shortage of machines for spreading fertilizers and much spreading has had to be by hand from trucks. On the other hand, the Soviet Union has made good use of aircraft for this and other agricultural processes.

AGRICULTURAL AVIATION

The Soviet Union was among the first countries to use aircraft for agricultural purposes, beginning with an attack on locusts in 1925. By 1932, 224 aeroplanes were being used for agricultural work.[2] After the war, as in other countries with wide expanses of land and large farms, it was found that, in addition to crop protection, the application of mineral fertilizer from the air was both practical and economic and this use has grown greatly in recent years. (Table 27).

The figures reflect the major increase that took place in the agricultural use of aircraft heralded in the government announcement at the Party Plenum in February, 1964. As early as 1948 a specialized aeroplane, the Antonov AN–2, had been produced to conform with specifications issued by the agriculture and forest authorities and more than 5,000 of these aircraft were built between 1949 and 1962, many of them being

[1] *Izvestiya*, 12.3.67, 2.
[2] Brezhnev and Minkevich (1958) 1961, 80.

exported. An improved version, the AN–2M, introduced in 1964, is powered by a 1,000 h.p. engine and carries a chemical payload of 1,500 kg. Other aircraft have been developed for agricultural work, including the MI–1 helicopter, while a

TABLE 27

THE GROWTH OF AGRICULTURAL AVIATION WORK IN THE U.S.S.R.

Type of work	*Area treated (million hectares)*				
	1940	*1950*	*1955*	*1960*	*1965*
Protection against crop diseases and pests	0·9	2·6	6·2	13·7	26·8
Weed control, etc.	—	0·1	0·6	2·3	11·4
Fertilizer application	—	0·7	3·1	4·1	16·8

Source: Slavkov and Tyutyunnik (1968), 4.

specialized version of the newer MI–4, the MI–4S, has been equipped with a hopper to carry 1,000 kg. of chemicals.[1] The AN–2, however, remains by far the most important agricultural aircraft (Table 28).

TABLE 28

SOVIET AGRICULTURAL AVIATION WORK BY TYPE OF AIRCRAFT, 1965

	Distribution of fertilizer—area treated		*Crop protection—area treated*	
	thousand hectares	*per cent.*	*thousand hectares*	*per cent.*
AN-2	14,701·1	87·7	23,278·8	87·0
YAK-12	1,850·3	11·0	2,791·9	10·4
MI-1 helicopter	210·6	1·3	596·5	2·3
KA-15 helicopter	0·5	—	91·5	0·3
TOTAL	16,762·5	100·0	26,758·8	100·0

Source: Slavkov and Tyutyunnik (1968), 68, 86.

The relative importance of topdressing and chemical spraying varies in different areas with the former more important in the north and west. Availability of fertilizers as well as the needs of leached and heavily cropped soils influence the pattern. In Central Asia, Kazakhstan and Transcaucasia, crop protection is more important. (Table 29).

[1] Taylor (ed.) (1970), 479, 499.

The advantages of using aircraft for crop dusting and spraying include speed and efficiency of application, avoidance of long hauls of heavy and bulky chemicals, economy of labour and making possible attention to areas such as mountain pastures inaccessible to surface machines. Cost, however, is important and, before calling on *Aeroflot*, the monopoly contractor for aviation services, the agronomists must be satisfied,

TABLE 29

DISTRIBUTION OF AGRICULTURAL AVIATION WORK BY REPUBLICS, 1965

	Distribution of fertilizer—area treated		*Crop protection—area treated*	
	thousand hectares	*per cent.*	*thousand hectares*	*per cent.*
Azerbaydzhan	191·6	1·1	1,837·6	6·9
Armenia	0·2	—	217·4	0·8
Belorussia	433·8	2·6	237·0	0·9
Georgia	79·0	0·5	133·7	0·5
Kazakhstan	800·1	4·8	3,837·2	14·3
Kirgizia	155·8	0·9	261·0	1·0
Latvia	62·3	0·4	9·8	—
Lithuania	77·9	0·5	7·7	—
Estonia	56·3	0·3	0·4	—
Moldavia	275·3	1·6	308·9	1·1
R.S.F.S.R.	9,717·3	58·0	11,140·8	41·7
Tadzhikistan	47·1	0·3	1,180·7	4·2
Turkmenistan	40·8	0·2	223·1	0·8
Uzbekistan	306·6	1·8	2,485·0	9·3
Ukraine	4,518·4	27·0	4,950·5	18·5
TOTAL	16,762·5	100·0	26,758·8	100·0

Source: Slavkov and Tyutyunnik (1968), 68, 86.

unless there are emergency considerations such as saving a crop from pests or diseases, that aerial work is reasonably competitive with other methods in costs.

Application of fertilizers is the field in which there is most need to compare the costs of ground and air spreading. Table 30 shows that the rate charged for aerial topdressing may make this rather more expensive than using ground machinery but the difference is not great. Furthermore, these are average figures and under certain conditions, such as a lighter spread or the use of more concentrated fertilizer, the aerial application may be the cheapest.[1] In the last resort, the cost-benefit ratio may depend on the alternative demands for the farm labour and the availability of machines.

TABLE 30

COMPARATIVE COSTS OF AIR AND GROUND SPREADING OF MINERAL FERTILIZER (at the rate of 183 kg./ha.)

	Cost per hectare (rubles)				
	AN–2	*YAK–12*	*Helicopter MI–1*	*Fertilizer spreader (surface)*	*Hand application*
Aviation tariff per hectare dressed	1·60	1·60	3·20	—	—
Cost of ground spreading	—	—	—	0·63	1·18
Servicing personnel	0·16	0·27	0·24	0·33	—
Overhead costs	0·10	0·16	0·16	0·20	0·72
TOTAL COSTS	1·86	2·03	3·60	1·16	1·90

Source: Slavkov and Tyutyunnik (1968), 74.

DEVELOPMENT THROUGH WATER CONTROL

The effects of an adverse water balance are felt over the greater part of the U.S.S.R., most areas having either a surplus or a deficiency of water in the growing period. Hence, water control in all its aspects is most important.

DRAINAGE

It has been calculated that over one million square miles could be improved for agriculture in the Soviet Union by drainage. Permanently waterlogged soils include peaty and marshy soils, while seasonally affected soils of the gley, gley podzol and turf-gley types may be classed as waterlogged mineral soils. The water régime of these latter may fluctuate sharply from excessive to insufficient soil moisture with variations in precipitation from one season to another and one year to another, but, unless the duration of waterlogging is normally brief, such soils repay artificial drainage. Spring waterlogging

[1] Slavkov and Tyutyunnik (1968), 73–74.

delays cultivation and so shortens the effective vegetative period and lowers yields, while waterlogging in autumn hinders harvesting and autumn ploughing.

Between 1956 and 1965 the area of artificially drained land in the U.S.S.R. was extended from 8·4 to 10·6 million hectares. In the middle 'sixties the speed with which drainage schemes were effected was considerably increased, following the swing from development of the virgin lands to more intensive use of western lands of low productivity. This is shown in the following table:

TABLE 31

COMPLETION OF DRAINAGE OF LAND IN THE U.S.S.R.
(thousand hectares)

	1950	*1960*	*1965*	*1969*
R.S.F.S.R.	87	112	165	220
Belorussia	29	89	198	191
Ukraine	6	79	119	109
Lithuania	9	83	99	131
Latvia	31	74	78	84
Estonia	8	29	34	45
Moldavia	3	3	7	1
Georgia	11	2	4	14
Azerbaydzhan	—	—	4	—
Armenia	—	1	—	—
U.S.S.R. TOTAL	184	472	708	795

Source: N.kh. SSSR 1969, 500.

This table reflects the continuing importance of work in the Pripyat marshes of Belorussia and adjacent areas of the R.S.F.S.R., the Ukraine, and the Baltic republics. Important also in the R.S.F.S.R. improvements are western and central Siberian areas.

Substantial tracts of marshland occur also in the central parts of European Russia, near Moscow, throughout the area from Ryazan and Vladimir northwards to Novgorod and Leningrad, and between the Belorussian marshes and Leningrad, with smaller patches around Smolensk and elsewhere in this morainic and outwash zone.

Surface ditches and underground drains frequently feed reservoirs which are used when drier conditions prevail, and on

TABLE 32

USE OF DRAINED LANDS IN AGRICULTURAL PRODUCTION ON COLLECTIVE AND STATE FARMS, 1968
(thousand hectares)

	Total drained land in use	*Arable land (sown or clean fallow)*	*Sown area*	*Principal crops*				*Orchards, small fruits, etc.*	*Hay and pasture lands*	*Personal plots of collective farmers and workers*
				Grains	*Industrial crops*	*Potatoes and vegetables*	*Fodder crops*			
U.S.S.R. TOTAL	5,906·2	2,937·7	2,878·6	1,211·1	187·8	191·5	1,288·2	36·8	2,841·4	90·3
R.S.F.S.R.	1,656·6	690·5	669·3	253·6	63·4	36·9	315·4	4·3	951·3	10·5
Ukraine	1,134·7	449·1	448·5	191·8	49·3	37·6	169·8	5·2	661·8	18·6
Belorussia	1,109·4	519·9	516·6	197·2	26·5	63·0	229·9	0·3	585·9	3·3
Georgia	84·3	40·0	36·2	20·0	2·4	0·9	12·9	11·9	22·7	9·7
Lithuania	995·8	699·5	679·6	297·2	28·9	22·4	331·1	5·6	256·5	34·2
Moldavia	26·5	15·7	15·7	4·1	2·6	1·6	7·4	3·2	7·5	0·1
Latvia	599·7	366·6	358·8	181·1	14·3	20·6	142·8	5·6	217·2	10·3
Estonia	299·2	156·4	153·9	66·1	0·4	8·5	78·9	0·7	138·5	3·6

Source: N.kh. SSSR 1968, 392.

more intensively used lands the expense of pumping away drainage water is justified by the higher returns obtained.

The use of lands with a drainage network is given in Table 32 for collective and state farms.

The most comprehensive regional drainage scheme in Russian territory is that of the Polesye in Belorussia. The drainage of this extremely backward area of fluvio-glacial sands and clays, commenced in 1872, had produced by 1897 some 5,000 km. of canals, had converted nearly 400,000 hectares of former swamp to water meadow and about 130,000 hectares to arable uses, besides rendering about 1,300,000 hectares of forest land accessible. The value of improved land rose on average by 600 per cent., from four to twenty-eight rubles per dessiatin.[1] In general, however, the area remained poor and suffered devastation, including the destruction of the drainage works, in the First World War. In the part that became Polish territory little development occurred in the inter-war years, but the Soviet administration organized 'reclamation co-operatives' and encouraged other collective and state farms to increase drainage, a policy extended to the western area after the reunion in 1940. By the outbreak of the Second World War the total area of reclaimed land had risen to about 260,000 hectares. Further plans were interrupted by the war but in 1952 it was estimated that 3,500,000 hectares needed reclamation in Belorussia, with about another million in the Ukraine.[2] Work has since been pursued steadily in the area but much still remains to be done to transform this region of swamps and backward agriculture into a fertile grass and arable farming area.

The nineteenth-century 'expeditions' which achieved such remarkable results in the Polesye also carried out drainage in other Russian areas, including the reclamation of over one hundred separate swamp areas, despite apathy and opposition from some landowners who thought that the work would reduce the rainfall, others who thought it would spoil the fishing, millers whose dams were affected and other interests. General Zhilinskiy, the architect of these works, was later transferred to Siberia, and there, between 1896 and 1914, constructed some

[1] French (1959), 176.
[2] French (1959), 179.

3,000 kilometres of canals in the swamps of the Baraba steppe, making much-needed land available for settlement.[1]

Except in Lithuania, Latvia and the much smaller drained areas of Moldavia and Georgia, pasture and hay comprise the dominant uses of the reclaimed land. In the U.S.S.R. as a whole, however, on drained areas arable uses equal pastoral, and of the arable uses, fodder crops about equal grain, with the expected predominance of grain in the Ukraine and Georgia. Grain and fodder uses are about equal in the Baltic republics and Belorussia. Fruits of various kinds are produced on about one-third of the Moldavian and one-seventh of the Georgian drained lands.

Most of the drained lands are in the north-west, namely the Baltic republics and neighbouring oblasts of the R.S.F.S.R., Belorussia and the north-west Ukraine. A line following the parallel of 50 degrees N. from the Polish border to Kharkov and then north-west to the Kama to follow this river, and finally, the Northern Dvina, encloses most of these areas. Other important drainage projects have been completed in the west Siberian lowlands, notably Novosibirsk oblast, and in east Siberia near Irkutsk and Lake Baykal. In the Far East in the Amur valley, especially in the Jewish (Yevreyskaya) A.O. and neighbouring parts of Khabarovsk kray are other important drained areas. For maximum productivity in the east, irrigation facilities as well as drainage must be provided because of seasonal variation in moisture supply.

In southern regions, the main area requiring drainage is the Kolkhid lowland. Despite the high evaporation in Transcaucasia, the valleys are swampy because of low river gradients and high precipitation. Full utilization of the climatic advantages of this region requires a much more extensive drainage network. An important scheme, however, has drained large areas between the rivers Rioni and Khobi, and work has been extended into other areas.

AGRICULTURAL DEVELOPMENT OF PEAT BOGS

Improvement of peat bogs for agriculture was undertaken before the Revolution but has been greatly increased in recent years, with over one million hectares improved in Belorussia

[1] French (1963), 52–54.

alone. The Minsk experimental bog station was established in 1911 and its study of the Komarov bog has yielded valuable results.[1] This bog, in a depression of the Belorussian-Lithuanian upland, is underlain mainly by sandy deposits but natural drainage is seriously impeded. Initially the peat formed from mosses, common reed and bogbean, with birch, willow and sedge contributing later.

Experience here shows that the soil formation process is directly associated with the degree of drainage and the use of the bog soils. More intense drainage and usage accelerates compaction, mineralization and the humification of organic matter. Cultivation substantially changes the organic content, the conservation of which is essential for sustained productivity. It is recommended that shallow peat beds should be turned into meadows, but moderately thick peat beds with over 500 metric tons per hectare of organic matter should be used for cropping, in which grains are rotated with perennial grasses. Perennial grasses help to control wind erosion and to prevent fires, which accelerate loss of organic material. Nitrogenous fertilizers provide an inexpensive means to further improvement and sustained productivity of these soils. In spite of increasing availability of artificial fertilizers, grasses and grains rather than row crops such as potatoes, sugar beets and fodder beans are still recommended. These latter crops give maximum yield of dry matter and protein, but also cause maximum loss of nitrogen, and concentration on the crops which give better conservation of elements even at the expense of productivity is preferred.

IRRIGATION

Irrigation is essential in arid and semi-arid areas for intensive cultivation and greatly increases productivity wherever the water balance is inadequate or unreliable. Irrigation is the only means in water-deficient areas of ensuring adequate water to growing plants. Further, it alters the heat balance in the lower layers of the atmosphere, as evaporation lowers temperatures. The effect of this is seen most markedly in Central Asia, where, with sufficient irrigation in oases, only 8 per cent. of July sukhoveys cause damage to cotton plants, as against 60 per cent.

[1] Skoropanov (1968).

in the desert areas.[1] Increased humidity resulting from irrigation also reduces the amount of dust carried into the air when the soil becomes excessively dry. Not only is soil lost in this process but a dust haze may last for several days, and if repeated frequently in the summer adversely affects the ripening of crops.

Before the Revolution, about 3·5 million hectares were irrigated, mainly in Uzbekistan and elsewhere in Central Asia. Another three million hectares were added by 1950, and the rate of addition has since been increased to 300,000–400,000 hectares per year (1964–68). Almost exactly half of these recent additions are in the R.S.F.S.R. and the Ukraine, about two-fifths in Central Asia and the balance in Transcaucasia and Moldavia.[2]

Land with irrigation networks now measures about 5·5 per cent. of the total cultivated area of the U.S.S.R., or under 2·3 per cent. of the total agricultural area, but the value of the crops produced with irrigation is over 15 per cent. of the total value of agricultural production.[3] Irrigation of pasture lands results in the irrigated areas exceeding the arable area in Azerbaydzhan and Turkmenistan, but in all other republics the reverse is true.

Soviet statistics of irrigated areas include lands with irrigation systems lacking water supply or temporarily out of use through salt or sediment accumulations or other reasons. Areas of estuary irrigation are not, however, included.[4]

In 1968, of about 12 million hectares with irrigation networks, 10·1 million hectares were ready for use and connected with their water supply and 9·8 million hectares were in use. About 87 per cent. of the irrigated lands are in arid regions, 11 per cent. in steppe areas, nearly 2 per cent. in the wooded steppe and 0·3 per cent. in forest regions. The distribution of these areas among republics and among uses is shown in Table 33.

Of the Central Asian republics' total of nearly 6 million hectares of irrigated land in use, 5 million hectares were sown

[1] Borisov (1959) 1965, 182.
[2] *N.kh. SSSR 1965*, 527; *1968*, 518.
[3] *World Atlas of Agriculture*, I, 487. The relationship between agricultural, cultivated, arable and irrigated areas for each republic in 1949 and 1958 appears on p. 488 of this atlas.
[4] *N.kh. SSSR 1967*, 934.

to crops in 1968 and half of this cropped area carried industrial crops, mainly cotton. In Kazakhstan and Kirgizia, however, grain and root crops far exceeded industrial crops, which was true also of the irrigated areas elsewhere in the U.S.S.R.[1]

Of the R.S.F.S.R. total of 1,602,300 hectares of land actually irrigated in 1968, three-fifths (992,200 hectares) was in the

TABLE 33

USE OF IRRIGATED LANDS ON COLLECTIVE AND STATE FARMS, 1968
(thousand hectares)

	Total irrigated land prepared for use	*Total irrigated land in use*	*Arable land (cropped or clean fallow)*	*Orchards, vine-yards, etc.*	*Hay and pasture lands*	*Personal plots of collective farmers, etc.*
U.S.S.R.	10,102·8	9,789·7	8,040·6	981·0	301·0	467·1
R.S.F.S.R.	1,694·4	1,602·3	1,206·1	184·3	176·0	35·9
Ukraine	746·1	736·0	641·7	87·6	3·1	3·6
Uzbekistan	2,670·4	2,639·0	2,323·7	179·1	0·9	135·3
Kazakhstan	1,265·9	1,200·9	1,062·3	73·6	9·4	55·6
Georgia	327·9	323·1	158·8	84·2	29·3	50·8
Azerbaydzhan	1,113·1	1,046·2	797·4	154·0	38·8	56·0
Moldavia	96·9	94·4	63·9	29·8	0·4	0·3
Kirgizia	870·8	866·3	750·7	40·2	17·8	57·6
Tadzhikistan	504·5	477·6	383·1	54·3	10·4	29·8
Armenia	248·0	239·7	136·3	66·7	14·9	21·8
Turkmenistan	564·8	564·2	516·6	27·2	—	20·4

Source: N.kh. SSSR 1968, 389.

North Caucasus economic region, the balance mainly in the East Siberia and Volga regions. 326,300 hectares of the North Caucasian total was in Dagestan, nearly half of this being under grain (140,300 hectares), with further substantial areas under orchards and vineyards (68,800 hectares), fodder crops (57,500 hectares) and hay and pasture (16,800 hectares). Grain crops were dominant on the irrigated lands of Krasnodar kray (82,500 out of 142,600 hectares) but these came second to fodder crops in Stavropol kray and Rostov oblast. In East Siberia, hay and pasture occupied over half the irrigated land, but within the region sharp differences occurred, with grain and fodder crops

[1] *N.kh. SSSR 1968,* 389–390.

most important in Krasnoyarsk kray and Tuva A.S.S.R.[1] Thus, while cotton is dominant in the hottest irrigated areas, there is considerable variety of use in the U.S.S.R. as a whole. Virtually all cotton and rice crops are irrigated, and irrigation is applied to about 30 per cent. of the lucerne, 24 per cent. of market gardens and 18 per cent. of orchards and vineyards.

Canals and dams harnessing the great rivers of the warmer regions of the Soviet Union have been of critical importance to the extension of the irrigation systems. Half the irrigated land in Central Asia draws water from the Amu Darya and Syr Darya rivers. The mountain headstreams of the Syr Darya have been intercepted by the North and Great Fergana canals, which together encircle the whole Fergana basin and supply the serozems with water through a complex network of distribution canals, making this the principal cotton growing region of the U.S.S.R. Another major cotton area supplied by the Amu Darya is the reclaimed part of the Golodnaya (Hungry) steppe, west of the Fergana basin, and other schemes have been completed or are under construction on this river. The Amu Darya has a greater flow than the Syr Darya but also carries a heavy load of silt, which hinders its utilization. It has been harnessed more gradually than the Syr Darya but supplies several irrigation networks and the Kara Kum canal, which carries much-needed water to southern Turkmenistan.[2]

The large river schemes usually combine irrigation with development of hydro-electric power, flood control and, where relevant, improvement of navigation, but local schemes have also been implemented. In Kazakhstan, ancient irrigation works have been reconstructed, and fields have been enlarged to suit increased use of machinery.

Irrigation is, however, expensive, and though long-term results may handsomely repay expenditure, projects which give a quick return, or involve relatively little investment, are sought. Estuary irrigation in steppe regions thus attracts considerable attention. Protective levees in these areas require on average 5–8 cu.m. of earthworks per hectare, costing 15–25 rubles (i.e., about 3 rubles per cubic metre). Mineral and organic fertilizers

[1] *N.kh. RSFSR 1968*, 212–213.
[2] For a summary of the canal systems and dams of the U.S.S.R. see Mellor (1964), 191–193. For further developments to mid-1967, see Sheehy (1967).

in such irrigated areas have raised hay yields from 5 to 20 centners per hectare in Taldy-Kurgan oblast of the Kazakh republic, while 40 centners of hay are harvested in more favourable lands of this type in Volgograd oblast, and 80–100 centners is considered a reasonable target for such haylands.[1]

Conservation of snow and ice meltwater is critically important in the steppe areas. One dam, 300 m. long and 1·5 m. high, built to retain thaw water on an area of 380 hectares, is considered responsible for a yield of 4,500 centners of hay compared with 900 centners usually harvested on the area.[2] Such constructions are doubly economical because they are undertaken in winter when work on the land falls off.

DRY FARMING

Although irrigation is the most promising means of increasing agricultural output in arid areas, it cannot be applied to all such areas, and dry farming methods must continue. The essence of good dry farming is choice of crop rotation and control over grazing. In Kazakhstan, dry farming must be practised widely, and a suggested suitable rotation system is six years of perennial grasses, one year of cocurbit and millet crops and one of grains.

Choice of crops is important, both for the protection of the reclaimed land and for the yields available to offset the cost of improvement. On the sandy soils on which much conservation work has been done, winter rye is generally the most productive grain, and is used as hay, 'artificial grazing' and for fertilization. Millet is also hardy and yields well. Cocurbits, especially melons, yield highly on warm sands. With deep ploughing, wide-row sowing (40–50 cm.) and snow retention by windbreaks, crops in the semi-desert areas of Kazakhstan commonly yield 6–7 centners per hectare of barley or millet, 70–80 centners per hectare of water melons, or 10–12 centners per hectare of lucerne hay.[3] Some sandy areas have shown considerable potential for viticulture and orchard crops, notably in the Don, Dnepr, Terek and Volga riverine areas.

In dry areas such as the virgin lands, where winter precipitation provides about a third of the annual moisture supply,

[1] Badiryan, (1956) 1960, 101.
[2] Badiryan, (1956) 1960, 101.
[3] A. E. Ivanov in Albenskiy and Nikitin (1956) 1967, 319–320.

snow accumulation is most important to reinforce soil moisture against the late spring drought. Strong winds blow snow off the fields, so that any measure which reduces the effect of surface winds helps to stabilize the snow, and cultivation which does not bare the fields of stubble is accordingly beneficial. Such preservation of the stubble, it is claimed, could give grain yields two to four centners per hectare higher than with mouldboard ploughing, and in the autumn of 1966 such cultivation was carried out on about 12,500,000 hectares in Kazakhstan, Altay kray and Omsk and Orenburg oblasts.[1]

Trench agriculture has been developed by the Aral experimental station. In a sandy area with an annual average of only 150 mm. of precipitation, numerous crops are grown, including berry and tree fruits and such high consumers of moisture as cabbage and cucumber. Potatoes have yielded 500–700 centners per hectare, tomatoes, 1,000–1,500 and water melons, 1,500–1,800, according to reports from farms in the Aktyubinsk and Kara-Bogaz-Gol areas. Trenches, dug down to the water table, are typically 1·5–2 m. deep, of similar width at the bottom and twice as wide at the top. They are filled with material from the upper layer of the soil.[2]

Further reference is made to dry farming methods in the following section.

INTENSIFICATION OF AGRICULTURE ON THE STEPPES & THE EROSION PROBLEM

The transformation of the steppes has occurred in two main stages, from pastoralism to extensive grain cultivation and from either of these to more intensive crop and livestock husbandry. It began in the nineteenth century with the southward spread of grain cultivation into the steppes of the Ukraine and Don from the wooded steppe and the beginning of the cultivation of the Siberian steppes. A second wave of this type of development came in the 1950s with the virgin lands scheme. The second stage overlapped the first, as in some regions more balanced forms of agriculture replaced extensive grain growing in the nineteenth century,

[1] A. Barayev, *Pravda*, 16.2.67, 2; *C.D.S.P.* 19, (7), 27–28.
[2] A. E. Ivanov, in Albenskiy and Nikitin (1956) 1967, 319–320.

and this second stage has already been reached in some of the virgin land development areas.

Soil erosion soon became a problem after the intensification of agriculture in the steppes, and though its ravages were not fully appreciated some farmers were already taking steps to control it early in the nineteenth century. Growing realization that tree planting is the simplest and cheapest method of reducing erosion and improving the hydrological regime stimulated various experiments, but there was little coordination. Government agencies had occasionally attempted to carry out afforestation projects in the southern regions from the time of Peter I, but with little effect on the mass of landowners. Extending cultivation on to light and sandy soils frequently produced disastrous results. It is said that in the Lower Dnepr area, blowing sands buried over 10,000 hectares of cultivated land between 1843 and 1868. Over-exploited sandy soils in the Don region formed drifts which forced the relocation of four townships, three smaller settlements and 84 farmsteads in the late nineteenth and early twentieth centuries.[1]

In 1809, V. Lomikovskiy began planting trees along fields in the Mirgorod district, besides experimenting with various rotations and the adaptation of land use to variations in relief. He claimed that his 'tree-and-field husbandry' enabled him to obtain good harvests during the widespread crop failures of 1834 and 1835.[2]

The more scientific approach to the development of the steppes owes much to the pioneer research of V. V. Dokuchayev and his colleagues and successors working in the Special Expedition of the Forest Department from 1892 in the steppes between the Dnepr and the Volga. They experimented with the establishment of forest shelterbelts on field boundaries and gulleys, the damming of ravines, construction of irrigation systems and ponds and made detailed soil surveys.

Perhaps the most striking regional development begun under Dokuchayev was the improvement of the Kamennaya steppe (Voronezh region). Tree planting and research

[1] I. S. Matyuk, in Albenskiy and Nikitin (1956) 1967, 277.
[2] Albenskiy and Nikitin (1956) 1967, 12–13.

began in 1892 but were followed by the wartime and revolutionary years, during which little work was practicable. In the 1930s a new research programme was initiated at the Kamennaya steppe research station, with emphasis on the drought-tolerance of different crops and the effects of forest belts on the chernozem soils. The conclusion was reached that shelterbelts increased the coefficient of utilization of precipitation and infiltration, raised the water table and reserves of productive moisture in the soil and reduced surface runoff from melting snow.[1] This station is now incorporated in the Regional Agricultural Research Institute of the Central Chernozem Region and much wider research work is being conducted.

Planting continued locally in many areas during the more peaceful periods of the following decades, and after the Second World War Stalin introduced his grandiose 'Plan for the Transformation of Nature', of which afforestation and irrigation were major parts. The plan visualised the establishment of six great shelterbelts, each some 400 m. wide and hundreds of miles long from north to south across the steppes. Though the plan was abandoned after Stalin's death, valuable belts of young trees had been established, particularly in the more northerly areas, where conditions for tree growth were good. Further south, survival rate was low, but work continues on a less spectacular scale by farms and forest organizations.

It is now accepted that forest belts contribute to soil and water conservation in several ways—by protecting soil from wind erosion, reducing drifting of snow and loss of infiltration from melting snow and, sometimes at least, by improving the water balance through transpiration. Some Soviet scientists also argue that summer rainfall can be increased by 10–20 per cent. in the east of European U.S.S.R., by a system of wind-reducing shelterbelts. Experiments on the Caucasian steppes suggest that the value of shelterbelts is particularly marked for high yielding crops and does not decrease in winters with little snow.[2] Improvement of planting techniques has made possible the establishment of forest

[1] Godunov (1968) reviews the history of these developments and provides a useful bibliography of research into steppe conditions.

[2] Kakushkin (1967).

belts in extremely arid areas of Uzbekistan, and it is claimed that the costs are repaid in full within seven years of planting.[1]

SOIL EROSION

Soil erosion is most extensive in the steppes but also appears over vast areas of the territory of the U.S.S.R., where forest cover has been destroyed and cultivation and grazing introduced. The 'slash-and-burn' methods of clearing forest and preparing land for shifting agriculture began the process of large scale tree-felling, later intensified as population pressure increased and markets for timber attracted commercial exploitation. The records of royal, monastic and private estates show that in the sixteenth century peasants were encouraged by tax concessions to clear land, while, as the pioneer fringe of settlement moved southward through the forest-steppe zone, much clearance was undertaken for defence works.[2] Demands for firewood, shipbuilding, iron smelting and potash processing were increased by voluminous exports of timber.[3] Much clearance was necessary before potential agricultural land could be made productive but, as in many other countries, and even in the present century when more enlightened practices might be expected, clear felling and complete lack of conservation practices totally destroyed timber resources in many areas and exposed the soil to accelerated erosion.

The three-field system offered little protection to the land, which was ploughed year after year and, even when fallow, gained only the protection of the weeds that established themselves. Contour ploughing, even had the peasants or the commune appreciated its desirability, would have been impracticable because of the strip form of working, while demarcation ditches could initiate rills and gullies. The shortage of grazing resulted in excessive use of steep banks particularly liable to erosion.

An indication of how widespread the soil erosion problem is in the Soviet Union today is given by Figure 21. Almost all cultivated areas suffer to some extent. Erosion is not serious in the forest zones, though in more intensively cultivated

[1] Boiko (1967).
[2] French (1963), 45–46.
[3] French (1963), 47–50.

areas near Moscow, Leningrad, Minsk, and other cities, and in parts of the Baltic republics and the Ural mountains, between 5 and 20 per cent. of the agricultural land is estimated to be moderately or severely affected by water erosion. In addition, large areas of sandy soils, especially in Belorussia and the northern Ukraine, are particularly vulnerable if ploughed. In the Perm-Upper Kama areas of the Urals, high river terraces, mainly under the plough, are eroded to a point where soil conservation measures are essential.[1]

In the forest-steppe and steppe areas, the problem of erosion is more general, and in some parts acute. This occurs because of the nature of relief and the large area under the plough, although the chernozem soils are less liable to sheet erosion than are grey forest soils and podzols, because of their higher humus content and superior structure.[2] In the forest-steppe and some more northerly steppe areas, erosion by water is as important as wind erosion. Thus, heavy rainstorms are principally blamed for the erosion affecting over 20 per cent. of the area of the Moldavian republic. Gullies, ravines and river valleys, on whose sides water erosion is active, break up more than 30 per cent. of the cultivated land. In the central chernozem belt, notably the oblasts of Belgorod, Voronezh, Kursk and Lipetsk, over two million hectares, 15 per cent. of the agricultural area, suffer from water erosion, and ravines are stated to occupy 300,000 hectares.[3]

Kozmenko stresses the depth of dissection of land as an index of the danger from water erosion to agriculture, and, from an examination of hypsometric maps and geomorphological data, distinguishes five areas as particularly liable to erosion[4]. These are:

(1) An area in the middle of the Central Russian upland, including
  (*a*) forest-steppe areas around Orel and Tula and western parts of Ryazan and Tambov oblasts
  (*b*) the central chernozem area (Voronezh—Kursk—Belgorod)

[1] *S-g.z.* (1962) 1963, 92.
[2] Kozmenko (1956) 1967, 206.
[3] Skachkov (1967), 20.
[4] Kozmenko (1956) 1967, 207–208.

(2) the area of the Don ridge adjacent to the confluence of the Don and Chir rivers
(3) the Volga heights and right-bank area of the middle and lower Volga
(4) the Dnepr heights and right-bank area of the middle and lower Dnepr
(5) the Donets ridge

In all these areas, erosion affects up to 30 per cent. of the agricultural land, but the types of erosional processes differ according to the type of rock and other local variations. Thus in the central chernozem area, the Cretaceous rocks are brought to the surface readily by the plough and steep slopes are common, so that sheet erosion develops strongly.[1]

Wind is an important factor in soil erosion, especially in the southern steppes. Sukhovey and other strong winds create dust storms, which are particularly serious south of a line from Kishinev through Voronezh to Kuybyshev and over virtually all the Siberian steppes (Figure 21). Kazakhstan and western Siberia, Altay kray, the southern parts of Omsk and Novosibirsk oblasts are areas which suffer seriously from wind erosion. Further east also the problem is widespread. It was reported in 1963 that 82 per cent. of the arable land of the Khakass A.O. was suffering from wind erosion, with layers of soil 1–12 cm. deep being removed from ploughed land over significant areas.[2]

The combination of wind and water erosion is particularly serious in the most southerly areas of the U.S.S.R., where high evaporation causes soils to dry out, become powdery and acutely vulnerable to high winds and occasional torrential rainfall. In most of the foothill region of the Central Asian republics, over 40 per cent. of the agricultural land has been indicated as suffering from accelerated erosion (Figure 21). In the north Caucasus region, the figure is between 20 and 40 per cent. In Stavropol kray a quartering of harvest yields on severely eroded slopes and a 20–30 per cent. reduction on slightly eroded land have been reported. Here two-thirds of the precipitation occurs in summer and erosion is active

[1] Kozmenko (1956) 1967, 209.
[2] Skachkov (1967).

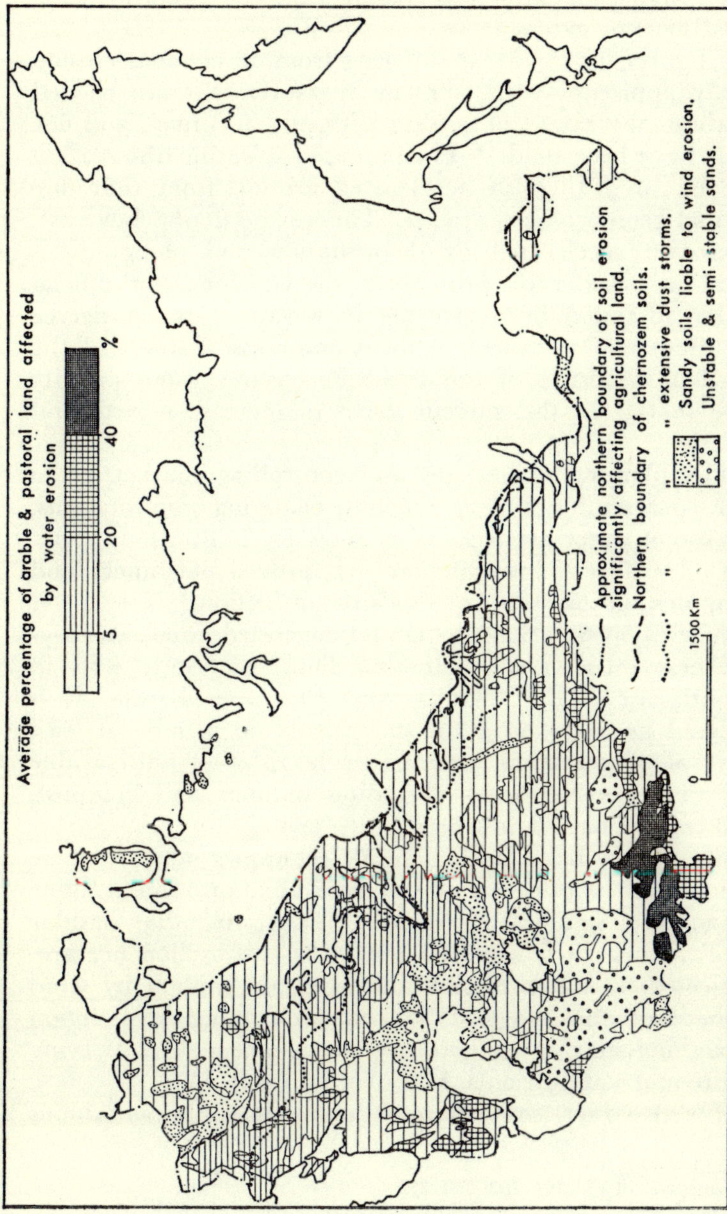

21. Soil erosion

Source: *Atlas sel'skogo khozyaystva S.S.S.R.*

on even slight slopes, no less than 1·5 million hectares being affected to some extent.[1]

The figures for the areas suffering from accelerated erosion are only approximations. Precise measurements are needed to establish the extent of erosion with any accuracy, and few of these have been made.[2] Furthermore, scientific observation is needed to distinguish accelerated erosion from that due to normal geomorphic processes. There is no doubt, however, that erosion accelerated by deforestation and ploughing is currently a considerable problem in the U.S.S.R., and official attention to it has been growing in recent years. A decree of the Central Committee of the Communist Party and the Council of Ministers of the U.S.S.R. ensures some priority for the matter in the current drive to increase agricultural production.[3]

Besides planting shelterbelts and controlling the cutting of timber, conservation measures include changing crop rotations, better use of water and control of snow for infiltration, application of fertilizers, introduction of improved machinery and development of improved methods of cultivation.[4]

Some erosion situations demand a concerted attack employing several interrelated methods. Thus, on sandy soils in the south-east of the Ukraine and European Russia, it is considered necessary to plant shelterbelts, to include at least four years of perennial grass cover in rotation with arable crops, to apply fertilizers, including manure and compost, and to replace bare fallow by a green ley.

While for the best results several techniques must often be combined, the application of one type of improvement alone may achieve significant results. Thus, in the Bashkir A.S.S.R., where it is reported that over two million hectares are affected by water erosion and over one million by wind erosion, observations suggest that simply changing the direction of ploughing may bring good dividends, through conserving moisture and soil nutrients, as shown in Table 34.

As noted earlier, soil erosion is severe in the mountainous

[1] Skachkov (1967).
[2] Kozmenko (1956) 1967, 207, 209–210.
[3] *Pravda*, 2.4.67, 1, 3; *Izvestiya*, 2.4.67, 1–2; *C.D.S.P.* 19, (13), 3–5.
[4] A convenient summary of standard erosion control methods in the U.S.S.R. is given in *World Atlas of Agriculture*, I, 488–489.

areas of Central Asia and the Caucasus. Cultivation here has had a very long history and there is much local understanding of the control of erosion by terracing and water control, but loss of soil has inevitably increased with use of steeper slopes for both cultivation and grazing. Much damage has resulted from overstocking and lack of fencing to control grazing, and deforestation has promoted accelerated erosion in many areas into which farming has extended.

TABLE 34

RELATIONSHIP BETWEEN GRAIN YIELDS AND DIRECTION OF PLOUGHING

Direction of ploughing	*Yield (centners per hectare)*			
	Spring wheat	*Winter rye*	*Oats*	*Permanent grass*
With slope (*vdol' sklona*)	9·5	15·0	10·5	23·3
Across slope (*poperek sklona*)	14·2	17·0	12·0	25·4

Source: Skachkov (1967).

On cultivated slopes, low walls of stones and debris are built to assist terrace formation. These should preferably follow the contours at intervals of 30–60 m., depending on the slope. On small, steep plots of dry-farmed land, belts of perennial grasses 10–15 m. wide should be established between fields for cultivation of 30–50 m. width.[1] On larger areas, belts of shrubs or trees and shrubs are recommended, spaced according to the gradient. Water diversion and drainage schemes are required in the sub-tropical, humid areas.

Tests conducted by the All-Union Scientific Research Institute of Tea and Sub-tropical Cropping are claimed to have shown that erosion may be practically halted by continuous shallow tillage and the distribution of tea trellises along the contours. Increasing the density of plants from under 10,000 per hectare to 80,000 or even to over 500,000 per hectare increased yields (8-year averages) from 4,905 kg. per hectare to 5,663 kg. and 6,202 kg. per hectare at the densities quoted, and improved protection against erosion.[2]

For citrus plantations, the Institute recommends that on

[1] Braude (1956) 1967, 270.
[2] Braude (1956) 1967, 271.

gradients of up to 15 degrees on krasnozem and weakly podzolic soils, safe cultivation without terracing is possible if the soils are ploughed deeply (45–50 cm.) but on steeper slopes terraces are desirable.

To restore pastures on steep, but cultivable, slopes, it recommends ploughing and sowing of contour strips, with fertilizers, and cultivation of the intervening strips when the new turf has become consolidated.[1] On other pastures, simply withdrawing stock and sowing grasses with appropriate fertilizers may effect recovery.

Avoidance of overgrazing is the principal conservation precaution on semi-desert pastures. On the Astrakhan sands at least 10 hectares per head of cattle and 2·5 hectares per head of sheep are recommended. To encourage uniform use of pasture lands and to prevent deterioration around wells, watering places should not be more than 6–7 km. apart. On flat sandy land used for winter pasture, the radius of grazing around a well may be increased to 6–8 km., but on a humpy or barchan type of sands, the radius should be reduced by 30–40 per cent.[2] Sown grasses and fodder crops may greatly increase carrying capacity without increasing the erosion hazard if the land is properly managed.

Conservation measures should be preceded by adequate survey and appraisal of the lands concerned. An example was the survey of soil cover, vegetation, moisture régime, relief and the results of experimental work on the sands of the lower Don valley.[3] The lands were classified into five groups, viz.:

Group 1. Lands of flat or rolling relief suitable for field cropping and occupying 25–30 per cent. of the tract. The soils are mainly leached sandy soils with chernozem characteristics.

Group 2. Hummocky land on the second terrace occupying 20–25 per cent. of the tract. This is regarded as unsuitable for agriculture, but suitable for tree plantations.

Group 3. Land between the first and second terraces,

[1] Braude (1956) 1967, 271.
[2] A. E. Ivanov in Albenskiy and Nikitin (1956) 1967, 321.
[3] A. E. Ivanov in Albenskiy and Nikitin (1956) 1967, 315–316.

occupying 4–6 per cent. of the area, regarded as suitable for orchards and vineyards. The soils are sandy loams with the water table 1–4 m. deep.

GROUP 4. Hummocky, overgrown, sandy areas, accounting for 30–35 per cent., of the lower Don lands, usable only for grazing.

GROUP 5. Land within the floodplains on the first (alluvial) terrace, with meadow and meadow-alluvial soils, appropriate to vegetable and fodder crops with irrigation.

Nation-wide application of conservation measures requires national systematization of survey methods. A framework of six levels of regionalization from phytoclimatic zones down to local areas, with appropriately increasing attention to the detail of conservation measures has been proposed by Silvestrov.[1] With the large numbers of organizations and individuals involved in conservation and the large area concerned, coordination on a spatial pattern is a matter of urgency, but the allocation of responsibilities by the C.P.S.U. Central Committee and U.S.S.R. Council of Ministers makes little reference to regional control and does not refer directly to the recognition of geographical, ecological and soil erosion units. The State Land Resources Institute is, however, charged with duties which could include such a national responsibility as shown, together with the responsibilities of other bodies in the conservation drive, in Table 35.

Besides the bodies named, which carry the main burden of the work on a national and republic scale, all agricultural, soil, forestry and water institutes must devote an increasing proportion of their effort to the conservation drive. The press, radio, television, films and other media are used for dissemination of propaganda on the vital nature of the work.

The U.S.S.R. State Bank is authorized to grant credits to collective farms for the conservation work they undertake, and the necessary acquisition of machinery. Personal responsibility for implementing the work, maintaining forest belts and correctly using the land is placed on the leaders and specialists of the farms and other organizations charged with the various conservation tasks.

[1] Silvestrov (1965).

TABLE 35

ALLOCATION OF RESPONSIBILITIES FOR SOIL CONSERVATION

U.S.S.R. Ministry of Agriculture and Union-republic Councils of Ministers	Annual programmes in soil conservation, budgets, work force, etc. required. Allocations of funds for control work on state farms by agricultural agencies.
U.S.S.R. Ministry of Finance	Annual allocations of funds to agricultural agencies for conservation work on collective farms, and to forestry agencies for soil conservation work.
U.S.S.R. Ministry of Agriculture	Implementation of erosion control measures on state and collective farms and other state lands. Establishment of state inspection service for soil conservation.
State Institute of Water Resources Research and Design, All-Union State Forestry Design Institute in collaboration with the U.S.S.R. Ministry of Agriculture, land use planning and other research institutes.	Planning and organization of territories for application of control measures to entire river basins without undue dispersal of funds on uncoordinated projects
State Land Resources Institute	Scientific questions on rational use of land resources and regulation of land use, registration and evaluation of soils and prediction of their use.
U.S.S.R. Ministry of Land Reclamation of Water Resources, All-Union Farm Machinery Association and State Forestry Committee.	Performance of work including terracing, flood control works, shelterbelt creation, planting of gullies and unproductive land.
Ministry of Tractor and Farm Machine Building	Design of new machines and tools for conservation work.
U.S.S.R. Council of Ministers' State Committee for Vocational and Technical Education and Union-republic Councils	Expansion of rural vocational and technical schools and their organization, special training in reclamation and associated mechanization.

Source: Adapted from articles in *Pravda,* and *Izvestiya,* 2.4.67.

Special attention is being given to the former virgin land and other steppe areas of Kazakhstan and west Siberia, where rapid development of erosion followed the ploughing campaigns.

For the U.S.S.R. it was proposed that between 1968 and 1970 forest shelterbelts be planted on an area of 324,000 hectares; gullies, ravines, sandy lands and other unproductive areas amounting to 827,000 hectares be afforested; 89,000 hectares of steep slopes be terraced; and erosion and flood control works costing about 188 million rubles be undertaken. This would merely represent the preliminary works in the vast drive, embodied in the new national scheme, to bring erosion under control.

The need to develop agriculture under very difficult physical conditions stimulated attention in the Soviet Union to methods of land evaluation. The assessment of potential for cropping and grazing has been taken beyond the work in agricultural climatology and soil analysis already described in attempts to assess the complete physical environment for agriculture. Attention to the economic aspects of land evaluation has been more recent but has been emphasized increasingly as the high cost of development of marginal lands has been realized, and as increased effect has been given to pricing systems which seek to offer comparable rewards to producers in varying physical conditions.[1] In the absence of a market in land and of land rent, survey of production potential assumes a vital role in the establishment of price zones. Geographers have been prominent in organizing discussion on the need for a land cadaster (hitherto rejected as unnecessary in a socialist society) to facilitate rational classification of land according to its fertility and ease of cultivation.[2] Many schemes for classification have been proposed,[3] but there has, as yet, been little evidence of large-scale mapping of substantial areas on a uniform basis.

[1] Jensen (1968).
[2] Zvorykin (1965) 1968, 166.
[3] See articles and bibliography in the issue of *Soviet Geography* (vol. 9 no. 3 1968) containing the above articles.

CHAPTER 12

Conclusions

The geographical pattern of an industry is never quite static, and in the Soviet Union, a country of rapid economic development promoted by revolutionary theories and central control of resources, change has been a continuous and sometimes violent process. In the preceding chapters change has been stressed by the use of statistical reports which, though crude and imprecise, illustrate the directions of agricultural development in the U.S.S.R. since the recovery from the self-imposed national purge of collectivization and the struggle for survival in total war. The slow changes of the preceding centuries, during which Russia was held back by isolation and lethargy, coupled with serfdom and repression, contrast with the drama of the last few decades.

The twin background themes of agrarian evolution—the essential elements of the social and political struggle, and the physical setting of agriculture in the Soviet territories—throw into prominence the appalling difficulties which have faced the modernizing agriculturalists. They have had to contend with a rural population of largely uneducated and apathetic peasants, depleted of its more enterprising men and women by the attractions of city life as well as by the indiscriminate toll of warfare. Economically, the agriculturalists long suffered the constant erosion of hard-earned gains as the state wrung their produce from them at minimum prices to subsidize industrial progress. Politically, they were at the mercy of central and local administrators and Party directives which seemed perversely calculated to make the job of the practical farmers even harder than nature made it. Up to the time of writing, this great labour force, with 224 million hectares of arable land and 374 million hectares of hay and pasture land at its disposal, has been unable to do more than slowly improve food production to a level at which it barely meets the needs of the Soviet population. Hunger and the fear of starvation are no longer part of the Russian scene, as they were even early in the present century, but the agricultural industry has failed, as yet, to provide anything approaching the variety of food available to the advanced 'western' nations or substantial amounts of any foodstuffs or raw materials for export.

As a result of this situation, unsatisfactory primarily to the Soviet governments and people, there has been unlimited scope for criticism of Soviet agriculture. Within the U.S.S.R., criticism has taken the form normal to a society in which the fundamentals of state organization are not open to question. Argument revolves around the details of organization and practice, with little or no challenge to the official assumption that major change can only be in the direction of a greater degree of socialization of the means of production, such as the replacement of collective farms by state farms, a return of land to private enterprise being unthinkable. In the western world, however, criticism has stemmed from the equally deeply implanted ideological devotion to private enterprise (supported generously by state subventions) characteristic of those countries. Western writings on the subject of Soviet agriculture have been overwhelmingly condemnatory of the collective system and have been preoccupied with the production failures of the system and the sufferings of the peasants in their huge, impersonal and state-directed farms. The reader of these depressing accounts of the system and its consequences could be forgiven if he concluded that we could write off the Soviet experiment in community farming as irrelevant to the rest of the world—at least to the non-communist world—and unlikely ever to become an efficient system.

Developing nations may, however, find rather more interest in the role forced on Soviet agriculture during the industrialization drive, as in the view expressed by Wilber:

> Soviet agriculture succeeded in fulfilling its two major functions of providing a growing marketed surplus of agricultural products and of freeing and utilizing surplus agricultural labour. The collective organization of agriculture was the key to this success. Increases in output per man and in man days worked were effected through

land reorganization and mechanization. These increases enlarged the potential marketable surplus (which was then taken by the state for investment in industry) and freed the labour necessary for the expanding industrial sector.[1]

Wilber compares the growth rate of agricultural production in the U.S.S.R. from 1928 to 1965 with the rates recorded by other countries at comparable stages in their development.

TABLE 36

TOTAL OUTPUT OF AGRICULTURAL PRODUCTS AND OUTPUT PER HEAD OF THE POPULATION

	Annual averages of output (million metric tons)				*Output per head (kg.)*			
	1909–13	*1946–50*	*1961–65*	*1966–68*	*1909–13*	*1946–50*	*1961–65*	*1966–68*
Grains	72·5	64·8	130·3	161·8	450	360	570	690
Potatoes	30·6	80·7	81·6	95·2	190	450	360	407
Sugar beet	10·1	13·5	59·2	85·1	60	70	260	364
Meat	4·8	3·5	9·3	11·3	30	17	41	48
Milk	28·8	32·3	64·7	79·3	180	180	290	338
Cotton	0·7	2·3	5·0	6·0	4	13	22	26

Sources: N.kh. SSSR 1968, 7, 315, 318 and calculations therefrom.

He shows that Soviet agricultural expansion compares favourably with that of the U.S.A. between 1870 and 1900, and far exceeds the growth rate in the United States in later periods. It can rightly be argued that agricultural output in the U.S.A. has for some decades been held back by government measures to limit surpluses, but it must also be noted that the Soviet Union started in 1928 from a base of small, undercapitalized, inefficient peasant holdings operated by people with virtually no education.

There has been a marked improvement in total production of agricultural products per head of the population in recent years, as shown by Table 36, which gives averages of output for selected periods representing pre-revolutionary Russia, post-World War II, the early 1960s, and the late 'sixties.

It will be seen that for some years after the last war output of grains and meat *per capita* was below the pre-World War I

[1] Wilber (1969), 90–91.

level, and milk was no higher. All have now been increased above the 1909–13 average *per capita* by at least 50 per cent., while the industrial crops, sugar beet and cotton, have been multiplied several times. The output of potatoes has increased less than population has risen since the post-war period, 1946–50, according to these figures, but is over twice that of the pre-revolutionary period.

TABLE 37

CONTRIBUTIONS TO SOVIET AGRICULTURAL OUTPUT FROM (*a*) STATE FARMS AND (*b*) THE PERSONAL PLOTS OF COLLECTIVE FARMERS AND OTHER WORKERS

	(a) *Percentages from state farms*			(b) *Percentages from personal plots*		
	1940	*1950*	*1968*	*1940*	*1950*	*1968*
Grains	8	11	45	12	7	2
Potatoes	2	4	14	65	73	62
Sugar beet	4	3	8	6	—	—
Sunflower	2	5	18	11	4	2
Cotton	6	4	20	—	—	—
Vegetables	9	11	33	48	44	41
Meat	9	11	30	72	67	38
Milk	6	6	27	77	75	38
Eggs	2	2	26	94	89	60
Wool	12	12	41	39	21	20

Source: Adapted from *N.kh. SSSR 1968*, 321.

It must be noted that all production figures are estimates including commodities consumed on the farms. Soviet methods of assessing yields are open to criticism, as previously noted, but have been improved over the years and are reliable enough for period comparisons of this kind.

An important aspect of the increases in production of recent years is the improved contribution from the socialized sector. Whereas previously there seemed little hope of the dependence on small-scale subsidiary farming being broken, this has now been greatly reduced. The state farms are still increasing their share of total agricultural output (Table 37a) and this process may be expected to continue. The personal plot, though still important for supply of some commodities to the towns, has been reduced in importance as a source of food for the nation as a whole (Table 37b), and improved earnings in collective work have greatly reduced its financial importance to the individual kolkhoznik.

Two major trends are clear in the pattern of growth of production in the collectivized sector. First, there was the virgin land scheme, i.e. development on extensive lines; secondly intensification.

In 1950 the sown area in the U.S.S.R. was 146·3 million hectares, slightly less than before the Nazi invasion some ten years earlier. Khrushchev's virgin lands programme increased the sown area to 195·6 million hectares, an addition of nearly 50 million hectares, or more than one-third. Thus, in about five years, there was added to the cropland of the Soviet Union an area about twice as great as the total area of Great Britain. Some of this increase came from a reduction in fallowing but about four-fifths came from the ploughing of the virgin lands. The expansion continued slowly to a peak of 218·5 million hectares in 1963, the year in which Khrushchev was deposed. It was generally agreed that expansion of the ploughed area had been taken far enough—most foreign observers judged too far. The accrued fertility in the 'virgin' soils had been already depleted by five to ten years of exhaustive grain cropping, and soil erosion was a growing problem. Much of this land has now been put under rotations including fodder crops and grass.

During the period of expansion there were some violent fluctuations in harvests, particularly in the virgin lands area, but the programme did bring into the grain-producing areas land which commonly experienced weather conditions contrasting with those in the traditional grain areas of the Ukraine and southern Russia. Thus, the eastern and western areas complemented each other and since 1960 the former virgin lands have contributed annually between 30 and 60 per cent. of the grain purchased by the state, according to the yields of the harvests in the different regions.[1] Furthermore, the availability of grain in large quantities from the east enabled more diversification to be implemented in the west, particularly the growing of fodder crops to facilitate the increase in livestock

[1] *Strana Sovetov za 50 let,* 138–139.

needed to provide a higher standard of living in the western cities.

As a result of the regional fluctuations the overall yield per hectare of grain remained at about the same level from 1950 to 1957. Seven-year averages showed a slight improvement from 7·6 centners per hectare for the period centred on 1951 to 8·3 for the seven years centred on 1954. From 1958, however, yields remained above 10 centners per hectare until the bad harvest of 1963. By this time the emphasis was beginning to shift from reliance on new lands being brought under the plough to more intensive farming. Improvement was resumed and an average of 14 centners per hectare was claimed for 1968.[1]

Although this level of yield is no more than moderate, the oft-quoted comparisons of yields of grain in the U.S.S.R. and the U.S.A. are quite inappropriate because the only grain areas at all comparable climatically are the Soviet Union's best lands in the Ukraine and the most northerly areas of the U.S.A. Gale Johnson showed that even in the 1950s Soviet grain yields were between 70 and 99 per cent. of those achieved in North Dakota, South Dakota and Nebraska.[2]

Improvements in yields have, to no small extent, depended on better deliveries of materials such as fertilizers and machinery to the farms. These increased fairly steadily during the years of the virgin land schemes, but the expanding area under cultivation took up a large proportion of the increased deliveries. Since this time the area under cultivation has grown only slightly but the deliveries of fertilizers doubled between 1958 and 1964 and rose by a further 70 per cent. to 1969, while the cultivated area remained approximately constant. Thus, in 1969, there was nearly four times as much fertilizer per cultivated hectare as in 1958.

The increases in production achieved in the three-year period 1966–68 as compared with 1961–65 have already been noted and, as these were achieved without further increases in the sown area, it is obvious that some intensification of cropping has, in fact, taken place. Taking into account

[1] *N.kh. SSSR 1968*, 349.
[2] Johnson (1963), 226.

the greater increases achieved in livestock production without an increase in the agricultural area, it is clear that, in agriculture as a whole, there has been an appreciable amount of intensification during the past decade. It may well be that intensification has been proceeding at an increasing rate.

Although the drive to bring virgin land under cultivation has almost stopped, the reclamation of ill-drained land and the extension of irrigation are proceeding steadily and constitute important elements in the intensification of production. Since 1964 there has been a steadiness of growth in agricultural production in the U.S.S.R. that previously there had never been. Although the advance in some of these recent years has been slight there have not been the sharp reversals characteristic of earlier periods. In general, since the war, the pattern of total agricultural production has been that of irregular advance for three to five years, then a fall in a year of bad harvests amounting to a loss of 10 per cent. or so of gross production, followed by a recovery taking two years or so to reach the previous high point, a further advance for three years or so, then another fall. Since 1964 the falls have not been sufficient to undermine the overall advance. The value of agricultural production at constant prices was in 1966–68 nearly twice that of the 1951–55 average and 20 per cent. above the 1961–65 average.[1] While Soviet statistics may continue to be viewed with suspicion there does not appear to be any reason to doubt the general trend.

Furthermore, this improvement must be seen against the diminishing rate of increase in the population. During the first post-war decade the Soviet population increased rapidly with the restoration of family life, and with the reduction of deaths from natural causes as the benefits of medicine and hygiene were extended into the more remote parts of the Union. During the decade 1959–68, however, each year showed a lower all-Union birth rate than the preceding one. The death rate reached a minimum of 6·9 per 1,000 in 1964, since when there has been a small rise as the proportion of aged people in the population has risen. The rate of natural increase slowly diminished to 26 per 1,000 in 1968. Even the present modest birth rate is sustained mainly by high rates in the

[1] *N.kh. SSSR 1968*, 314–316.

Central Asian and some of the Caucasian republics and, with the death rates reduced to European levels, the highest rates of natural increase also are in these areas. It seems likely that as birth control practices spread into these areas the already present trend of diminishing birth rates in them will be accentuated. Throughout the Union, birth rates are lower in cities than in rural areas and each year shows an increase in the proportion of the total population recorded as urban.

Thus, the demands of a rapidly growing population for more food merely to keep the same level of nutrition are diminishing, and quality has become increasingly important. This is not to say that the Soviet Union will never again suffer a crisis from food shortages; with the particular mixture of climates with which the Soviet Union is endowed one would have to be bold indeed to forecast continuous sufficiency. But recurrent crises seem to be less likely, and the stabilization of the population facilitates the achievement of balance.

If, then, one looks further ahead, will the Soviet Union become an important net exporter of foodstuffs and agricultural raw materials? This cannot be predicted confidently, but neither can it be confidently stated that this is unlikely, much less impossible. It is not many years since the Soviet Union was importing large quantities of grain to meet its needs, but in most years now it has a moderate amount for export. Given continued encouragement by the better prices offered to farms in recent years and the improved geographical spread of production resulting from the reclamation of virgin lands, grain surpluses could become normal. Soviet oilseed production, particularly from sunflowers, is already regarded as a possible threat to world prices of fats. In the past, the livestock branches have been particularly weak, with low milk yields per cow, and poor meat yields, and they still have a long way to go to equal performances in advanced livestock producing countries but milk yields per cow have improved by 50 per cent. since 1950 in the country as a whole. The Soviet Union is the biggest producer of butter in the world.[1] At present there is little for export and immediate increases

[1] Production figures have shown over 1 million tons per annum since 1965, approximately twice that of any other country, *Dairy Produce*, Commonwealth Secretariat, London, 1969, 37.

in production will readily be taken up if prices are kept steady or lowered relatively to other commodities. But the Russians are not traditionally big consumers of butter and there are signs of consumer resistance to paying high prices now for a food they have not always enjoyed in the past. If the Soviet Union began to export large quantities of butter there could be a serious problem when the international market is saturated. Of course, if the Soviet Union could solve the problem, which has baffled the capitalist countries, of diverting food production into channels which would help the needy countries of the world, there need be no problem of surpluses. Unfortunately, if the U.S.S.R. produces any excess of agricultural products, they are most likely to be of the kind that will directly compete with the producers of temperate lands where there is already a problem of surpluses and low producer prices.

There may thus come a time, if excess produce of the collective farms is offered to world markets, when the western farmers will be wistful for the days of lagging production in Soviet agriculture. Of this there is, let it be stressed, no immediate sign, but it is hoped that the geographical treatment of the themes in this book will help to show that a change in this direction could occur quite quickly if the Soviet government should overcome its problems a little more effectively than in the past—and there are signs that this will be done.

Three main needs still exist in Soviet agriculture. The first is continuing intensification of production, especially of output per hectare in the more fertile western areas which also commonly have a labour surplus, and of output per man in the eastern areas. The second need is continuing improvement of the position of the rural worker compared with the urban worker, involving an increase in incentives to farm workers. The third need is an increase in confidence in the countryside, for which stability in farm structure, or no more than relatively slow and moderate changes, would appear to be important.

Continuing intensification of production would seem to be accepted by the Soviet government. Investment is now running at a higher rate than ever before in agriculture, and industries which are important suppliers of farming inputs, such as the chemical industry, are also receiving priority in investment. The Soviet farms are much better supplied with technical advice and background scientific research, for example in soil studies, agro-climatology and biology, than those of many other countries. Better availability of resources should enable this work to be capitalized for greater application.

The position of the rural worker has been improving, both in terms of income and availability of services, for example schools, hospitals, theatres, but the problem of providing or upgrading all these in rural areas throughout the U.S.S.R. is a large one and there is still a great drift from the land. To some extent this remains a measure of adjustment to increasing mechanization but too many of the brighter young people go to the towns for their tertiary education and never return because of the limited scope for advancement in rural areas.

As far as stability is concerned, the forecasts may be made that the collectives will be allowed to continue side-by-side with the state farms, though probably with steady diminution of their number; and that the farm workers will be allowed to keep their private plots because these forms of land-holding offer substantial practical advantages to the Soviet state, although this is no more than a reasonable guess. Provided that the Soviet government and Party cadres do not further antagonize the countryside, a continuing improvement in Soviet agricultural output, probably at an increasing rate for at least the next decade, may be predicted. Of these predictions, the last is probably the most important in a world of rapidly increasing population and uncertain food supplies.

Appendix I

ORGANIZATION OF AGRICULTURAL LAND IN THE U.S.S.R.
BY TENURE 1 NOVEMBER, 1969 (million hectares)

	Total land in farms etc.	*Total agricultural lands**	*of which,* Arable	Hay	Pasture
KOLKHOZ lands, total of which,	360·1	211·6	110·8	15·8	81·3
lands in collective use (including reserve, *goszemzapas* and forest organizations)	355·0	206·7	106·7	15·6	81·3
personal plots of kolkhoz workers	4·8	4·6	3·9	0·2	—
kolkhoz lands in personal use of workers, etc.	0·33	0·3	0·25	0·01	—
SOVKHOZ and other state farm lands (including reserve, *goszemzapas* and forest organizations)	684·3	331·4	110·1	24·4	192·8
Lands in personal use of workers, etc. (excluding plots on kolkhozes)	3·5	3·2	2·4	0·4	—
Total agricultural lands and lands held by farms	1,047·9	546·2	223·3	40·6	274·1
GOSZEMZAPAS and forest lands (not included above)	1,122·4	42·5	0·4	5·7	36·0
Other land tenures	57·2	19·4	0·6	1·2	17·3
TOTAL LAND AREA	2,227·5	608·1	224·3	47·5	327·4

* Includes arable, fallow, orchards, vineyards, hay and pasture but excludes reindeer pasture

Source: N.kh. SSSR 1969, 304.

Appendix II

Administrative Divisions of the U.S.S.R.

A.O.	Autonomous oblast. Division of a republic or (in the R.S.F.S.R., a kray) with limited local autonomy for a small minority national group.
A.S.S.R.	Autonomous Soviet Socialist Republic. An area within a Union republic designated as the homeland of an important minority group, with some degree of local autonomy, otherwise serving the same functions as an oblast.
Kray	An administrative division comparable to an oblast but generally larger in area. Originally a kray contained subordinate A.Os or N.Os, whereas an oblast did not. This distinction has now been lost as some oblasts contain autonomous areas and the Maritime kray does not.
N.O.	National okrug. A thinly populated area having a distinctive nationality group; subordinate to an oblast or kray.
Oblast	Basic administrative subdivision of a republic, generally named after its capital.
Rayon	A subdivision of an oblast, kray or autonomous area, which may itself be subdivided into rural (village soviet) and urban areas.
S.S.R.	Soviet Socialist Republic. Also known as Union Republics, there are fifteen of these major political divisions, the largest being the Russian Soviet Federated Socialist Republic [R.S.F.S.R.] or Russian republic.
Union Republic	See S.S.R.

Bibliography

References are arranged by author or editor and date, except for Soviet atlases and statistical handbooks which are listed before the main body of references. Translated works are given under the name of the author according to the transliterated system employed in this book, together with the date of the original publication and the original title, followed by the translation details. Articles in translated books and journals are, however, given only in the translated form for brevity. The following abbreviations are used:

A.A.A.G. *Annals of the Association of American Geographers.*
C.D.S.P. *Current Digest of the Soviet Press.*
I.P.S.T. *Israel Program for Scientific Translations.*
N.kh. *Narodnoye khozyaystvo* (as below)
S.G. *Soviet Geography: Review and Translation.*
S.kh. *Sel'skoye khozyaystvo* (as below).
V.S.N. *Vestnik Sel'skokhozyaystvennoy Nauki.*

* * *

SOVIET ATLASES

Atlas razvitiya khozyaystva i kul'tury SSSR (Atlas of the development of the economy and culture of the U.S.S.R.) Moscow, 1967.

Atlas sel'skogo khozyaystva SSSR (Atlas of agriculture of the U.S.S.R.) Moscow, 1960.

Atlas SSSR (Atlas of the U.S.S.R.) 2nd ed. Moscow, 1969.

Fiziko-geograficheskiy atlas mira (Physical-geographical atlas of the world) Moscow, 1964.

SOVIET STATISTICAL HANDBOOKS

Narodnoye khozyaystvo RSFSR v . . . godu, statisticheskiy ezhegodnik (National economy of the R.S.F.S.R. in . . . Statistical yearbook), Moscow.

Narodnoye khozyaystvo SSSR v . . . godu, statisticheskiy ezhegodnik (National economy of the U.S.S.R. in . . . Statistical yearbook), Moscow.

SSSR v tsifrakh v . . . godu (The U.S.S.R. in figures), Moscow.

Sel'skoye khozyaystvo SSSR, Statisticheskiy sbornik, (Agriculture of the U.S.S.R., Statistical handbook), Moscow, 1960.

Strana Sovetov za 50 let, (The land of the Soviets after 50 years), Moscow, 1967.

* * *

Agabeili, A. A. (1961) Dairy buffalo breeding, in Rostovtsev (ed.) 121–122.

Akademiya Nauk SSSR (1962) *Pochvenno-geograficheskoye rayonirovaniye SSSR* trans. by A. Gourevitch, *Soil-geographical zoning of the USSR (in relation to the agricultural usage of lands)* Jerusalem, I.P.S.T. 1963.

Aksenenok, G. A. (1959) *Voprosy kodifikatsiy zakonodatel'stva o kolkhozakh* (Questions in the codification of statutes on collective farms) Moscow.

Alampiyev, P. M. (1961) The objective basis of economic regionalization and its long-range prospects *S.G.* 11, (8), 64–74.

Aleksandrov, N. P. (ed.) (1968) *Razmeshcheniye i spetsializatsiya zemledeliya i zhivotnovodstva v tsentral'no-chernozemnoy zone* (Distribution and specialization of cultivation and livestock rearing in the central-chernozem zone), Moscow.

Al'benskiy, A. V. and Nikitin, P. D. (eds.) (1956) *Agrolesomelioratsiya*, Moscow, trans. by A. Gourevitch, *Handbook of afforestation and soil melioration*, Jerusalem, I.P.S.T., 1967.

Alpat'ev, A. M. (1950) Ratsional'noye ispol'sovaniye osadkov—osnova preodoleniya zasukhi (Efficient utilization of rainfall as a basis of drought control) in *Agroklimaticheskiye usloviya stepi Ukrainskoy SSR i puti ikh uluchsheniya* (Agroclimatic conditions of the Ukrainian steppe and their amelioration), Kiev.

Anderson, J. (1963) Commentary, on paper by Jasny in Laird (ed.) 248–265.

Anderson, J. (1967a) A historical-geographical perspective on Khrushchev's corn program, in Karcz (ed.), 104–134.

Anderson, J. (1967b) Fodder and livestock production in the Ukraine, *East Lakes Geographer*, 3, 29–46.

Armstrong, T. (1965) *Russian settlement in the North*, Cambridge, Cambridge Univ. Press.

Arutyunyan, Yu. V. (1963) *Sovetskoye krest'yanstvo v gody Velikoy Otechestvennoy Voyny* (The Soviet peasantry in the years of the Great Patriotic War), Moscow, Akademiya Nauk SSSR.

Arzumanyan, E. A. (1961) Interspecific hybridization in cattle breeding, in Rostovtsev (ed.) 134–138.

Babushkin, L. N. (1957) Climatic characteristics of the summer atmospheric droughts and sukhoveis in the cotton-growing area of Uzbekistan, in Dzerdzeevskiy (ed.) 54–58.

Badir'yan, G. (1956) Problems in location and development of forage reserves on collective farms, in Polyakova, (ed.) 93–109.

Baiburtsyan, A. A. (1961) *Novyy metod povysheniya productivnosti skota*, Moscow, trans. by A. Birron and Z. S. Cole, ed. by Z. S. Cole, *A new method of increasing livestock productivity*, Jerusalem, I.P.S.T., 1964.

Baranskiy, N. (1950) *Ekonomicheskaya geografiya SSSR*, Moscow.

Belov, F. (1956) *The history of a Soviet collective farm*, London, Routledge and Kegan Paul.

Bergson, A. and Kuznets, S. (eds.) (1963) *Economic trends in the Soviet Union*, Cambridge, Mass., Harvard Univ. Press.

Blum, J. (1961) *Lord and peasant in Russia from the ninth to the nineteenth century*, Princeton, Princeton Univ. Press.

Boiko, N. P. (1967) Zashchitnoye lesorazvedeniye na bogare (Protective forest management on non-irrigated land), *V.S.N.* 1967 (2), 60–67.

Borisov, A. A. (1959) *Klimaty SSSR*, Moscow, 2nd ed., trans. by R. A. Ledward, ed. by C. A. Halstead, *Climates of the U.S.S.R.*, Edinburgh, Oliver and Boyd, 1965.

Bornstein, M. (1969) The Soviet debate on agricultural price and procurement reforms, *Soviet Studies*, 21, 1–20.

Botman, K. S. (1968) Natural restoration of soil fertility on terraces of mountain slopes, *Soviet Soil Science* (*Pochvovedeniya*) 1968 (4), 518–524.

Braekhus, K. (1968) Some geographic aspects of Soviet agriculture, *Norsk geografisk tidsskrift*, 22, 39–55.

Braude, I. D. (1956) Erosion in mountain regions and control measures, in Al'benskiy and Nikitin (eds.), 247–276.

Brezhnev, D. D. and Minkevich, I. A. (1958) *Osnovnye dostizheniya sel'skokhozyaystvennoy nauki v SSSR*, Moscow, trans. by R. Farkash and M. Paenson, *Achievements of agricultural science in the U.S.S.R.*, I.P.S.T., Washington, 1961.

Budyko, M. I. (1956) *Teplovoy balans zemnoy poverkhnosti*, Leningrad, trans. by N. A. Stepanova, *The heat balance of the earth's surface*, U.S. Dept. of Commerce, Washington, 1958.

Burlakov, N. M. (1961) Development of beef-cattle husbandry and breeding of new early maturing beef-cattle breeds in the U.S.S.R., in Rostovtsev, (ed.) 75–80.

Burmantov, G. G. (1966) The formation of functional types of settlements in the southern tayga, *Doklady Instituta Geografii Sibiri i Dal'nego Vostoka*, trans. in *S.G.* 9 (2), 1968, 112–119.

Clarke, E. D. (1810) *Travels in various countries, Vol.* 1, *Russia, Tartary and Turkey* (*1800*), London, Cadell and Davies.

Clarke, R. A. (1969) Soviet agricultural reforms since Khrushchev, *Soviet Studies*, 20, 159–178.

Conquest, R. (1968) *Agricultural workers in the U.S.S.R.*, London, Bodley Head.

Cox, K. R. and Demko, G. J. (1967) Agrarian structure and peasant discontent in the Russian revolution of 1905, *East Lakes Geographer*, 3, 3–20.

Danilov, V. P. (ed.) (1963) *Ocherky istorii kollektivizatsii sel'skogo khozyaystva v soyuznykh respuplikakh* (Essays in the history of the collectivization of agriculture in the union republics), Moscow.

Davies, R. W. (1955) The investment-consumption controversy: practical aspects, *Soviet Studies*, 7, 59–74.

Davitaya, F. F. and Sapozhnikova, S. A. (1969) Agroclimatic studies in the U.S.S.R., *Bulletin of the American Meteorological Society*, 50 (2), 67–74.

Dibb, P. (1969) *Soviet agriculture since Khrushchev, an economic appraisal*, Canberra, Australian National University.

Dobrynin, V. P. (n.d. 1964?) *Nauchnye osnovy zhivotnovodstva* (Scientific principles of animal husbandry) Moscow.

Dorozhkin, I. (1967) Kartofel', (Potatoes) in Pannikov and others, 120–121.

Dovring, F. (1966) Soviet farm mechanization in perspective, *Slavic Review*, 25 (2), 287–302.

Dumont, R. (1954) *Economie agricole dans le monde*, Paris, trans. by D. Magnin, *Types of rural economy*, London, Methuen, 1957.

Dumont, R. (1964) *Sovkhoz, kolkhoz ou le problématique communisme*, Paris, Éditions du Seuil.

Dunin-Barkovskiy, L. V. (1967) The water problem in the deserts of the U.S.S.R., *Problemy osvoyeniya pustyn'*, 1967, No. 1, trans. in *S.G.* 9 (6), 1968, 458–468.

Durgin, F. A. (1964) Monetization and policy in Soviet agriculture since 1952, *Soviet Studies*, 15, 375–407.

Durgin, F. A. (1967) Comment, on paper by N. Nimitz in Karcz (ed.) 206–211.

Dzerdzeevskiy, B. L. (1957) *Sukhovei, ikh proiskhozhdeniye i bor'ba s nimi*, Moscow, trans. by I.P.S.T., *Sukhoveis and drought control*, Jerusalem, 1963.

Emmons, T. (1968) *The Russian landed gentry and the peasant emancipation of 1861*, Cambridge, Cambridge Univ. Press.

Evseev, P. K. (1957) The make-up of summer sukhoveis in the south-east of the European territory of the Soviet Union, in Dzerdzeevskiy (ed.), 100–110.

Ewald, E. (1968) Development and international importance of Soviet soil science, *Soviet Soil Science* (*Pochvovedeniye*) 1968 (2), 143–151.

Fel'dman, Ya. I. (1957) Definition of meteorological criteria for sukhovei by means of complex climatology, in Dzerdzeevskiy (ed.) 87–99.

Fishevskiy, Yu. K. (1969) *Ekonomicheskiye i sotsial'nye problemy razvitiya sel'skogo khozyaystva SSSR* (Economic and social problems of the development of agriculture in the U.S.S.R.), Moscow.

Florenskiy, A. (1967) Romanovskiye ovtsy (Romanov sheep), in Pannikov and others, 165–166.

Florinsky, M. T. (1953) *Russia, a history and an interpretation*, 2 vols., New York, Macmillan.

French, R. A. (1959) Drainage and economic development of Poles'ye, U.S.S.R., *Economic Geography*, 35, 172–180.

French, R. A, (1963) The making of the Russian landscape, *Advancement of Science*, 20, 44–56.

French, R. A. (1967) Contemporary landscape change in the U.S.S.R., in Steel, R. W. and Lawton, R., (eds.) *Liverpool Essays in Geography*, London, Longmans, 547–563.

Galitskiy, M. I., Danilov, C. K. and Korneyev, A. I. (1965) *Ekonomicheskaya geografiya transporta SSSR*, Moscow.

Gerasimov, I. P. (1967) Basic problems of the transformation of nature in Central Asia, *Problemy osvoyeniya pustyn'*, 1967, No. 5, 3–17, trans. in *S.G.* 9 (6), 1968, 444–458.

Godunov, I. B. (1968) On the seventy-fifth anniversary of the Kammenaya Steppe Area, *Soviet Soil Science* (*Pochvovedeniye*), 1968 (1), 132–142.

Goryunov, N., Petrunin, V. and Sirgel'bayev, K. (1967) Orosheniye risa v Kazakhstane (The irrigation of rice in Kazakhstan), *V.S.N.* 1967 (6), 51–58.

Gradov, M. (1967) Volen'yem krayu (In the reindeer region), in Pannikov (ed.) 176–177.

Grigor'yev, A. A. and Budyko, M. I. (1960) Classification of the climates of the U.S.S.R., *S.G.* 1, (5) 3–24.

Gsovski, V. (1948–49) *Soviet Civil Law*, 2 vols., Ann Arbor, Michigan Univ. Press.

Gvozdetskiy, N. A. (1960) The physical-geographic regionalization of the U.S.S.R. for agricultural purposes, *S.G.* 1 (9), 5–19.

Hubbard, L. E. (1939) *The economics of Soviet agriculture*, London, Macmillan.

Hultquist, W. E. (1967) Soviet sugar-beet production: some geographical aspects of agro-industrial coordination, in Karcz (ed.) 135–155.

Hunter, H. (1957) *Soviet transportation policy*, Cambridge, Mass., Harvard Univ. Press.

Hutchings, R. (1971) *Seasonal influences in Soviet industry*, London, Oxford Univ. Press.

Ignatenko, I. V. (1967) Agricultural reclamation of tundra soils, *Doklady Soil Science*, 1967 (13), 1831–1834 (supplement to *Soviet Soil Science*).

Ignatov, L. (1956) Planning the development of commonly-owned livestock on collective farms, in Polyakova, (ed.) 243–258.

Jackson, W. A. D. (1959) The Russian non-chernozem wheat base, *A.A.A.G.* 49, 97–109.

Jackson, W. A. D. (1961) The problem of Soviet agricultural regionalization, *Slavic Review*, 20, 656–678.

Jasny, N. (1949) *The socialized agriculture of the U.S.S.R.* Stanford, Stanford Univ. Press.

Jasny, N. (1960) Peasant-worker income relationships, *Soviet Studies*, 12, 14–22.

Jasny, N. (1963) Low- and high-yielding crops in the U.S.S.R., in Laird (ed.), 215–265.

Jensen, R. G. (1964) Soviet subtropical agriculture: a microcosm, *Geographical Review*, 54 (2), 185–202.

Jensen, R. G. (1967) The Soviet concept of agricultural regionalization and its development, in Karcz (ed.), 77–98.

Jensen, R. G. (1968) Land evaluation and regional pricing in the Soviet Union, *S.G.* 9 (3), 145–149, with bibliography, 149–153.

Jensen, R. G. (1969) Regionalization and price zonation in Soviet agricultural planning, *A.A.A.G.* 59, 324–347.

Johnson, D. G. (1963) Agricultural production, in Bergson and Kuznets, 203–234.

Joravsky, D. (1967) Ideology and progress in crop rotation, in Karcz (ed.), 156–172.

Kabysh, S. S. (1965) The permanent crisis in Soviet agriculture, in Laird and Crowley (eds.), 164–182.

Kahan, A. (1963) Soviet statistics of agricultural output, in Laird (ed.), 134–168.

Kakushkin, V. N. (1967) Agronomicheskoye vliyaniye lesnykh polos v bessnezhnye zimy (The agronomic effect of forest shelterbelts in snowless winters) *V.S.N.* 1967 (2), 52–59.

Karcz, J. F. (ed.) (1967) *Soviet and East European agriculture*, Berkeley and Los Angeles, California Univ. Press.

Karcz, J. F. (1967) Thoughts on the grain problem, *Soviet Studies*, 18, 399–434.

Khlebnikov, V. (1967) Razvitiye sel'skogo khozyaystva v novykh usloviyakh (The development of agriculture in new conditions) *Ekonomika Sel'skogo Khozyaystva* (1967) 2, 3–12.

Klatt, W. (1964) Output and utilization of foodstuffs in the Soviet Union, *Studies on the Soviet Union*, New Series, III-4, 99–109.

Kolesnikov, E. V. (1967) Sady na svetlo-serykh lesnykh pochvakh (Orchards on light-grey forest soils) *V.S.N.* 1967 (1), 97–100.

Kostennikov, V. M. (ed.) (1965) *Ekonomiko-geograficheskiye rayony SSSR* (Economic-geographic regions of the U.S.S.R.), Moscow.

Kovalev, S. A. (1964) Problems in the Soviet geography of rural settlement, *Geografiya naseleniya v SSSR*, Moscow, 131–143, trans. in *S.G.* 9 (8), 641–651.

Kozmenko, A. S. (1956) Soil erosion and its control, in Al'benskiy and Nikitin (eds.), 187–246.

Kulik, M. S. (1957) Sukhovei criteria, in Dzerdzeevskiy (ed.) 59–64.

Laird, R. D. (1958) *Collective farming in Russia: a political study of the Soviet kolkhozy*, Lawrence, Univ. of Kansas.

Laird, R. D., Sharp, D. E. and Sturtevant, R. (1960) *The rise and fall of the MTS as an instrument of Soviet rule*, Lawrence, Univ. of Kansas.

Laird, R. D. (ed.) (1963) *Soviet agricultural and peasant affairs*, Lawrence, Kansas, Univ. of Kansas Press.

Laird, R. D. and Crowley, L. (eds.) (1965) *Soviet agriculture; the permanent crisis*, New York, Praeger.

Lemeshev, M. Y. (ed.) (1965) *Ekonomicheskoye obosnovaniye struktury sel'skokhozyaystvennogo proizvodstva*, (The economic basis of the structure of agricultural production), Moscow.

Leonov, I. (1967) Na tabachnykh plantatsiyakh, (On the tobacco plantations) in Pannikov and others, 123–124.

Lewin, M. (1965) The immediate background of Soviet collectivization, *Soviet Studies*, 17, 162–197.

Lewin, M. (1966) *La paysannerie et le pouvoir sovietique*, Paris, Mouton, trans. by I. Nove with J. Biggart, *Russian peasants and Soviet power: a study of collectivization*, London, Allen and Unwin, 1968.

Litoshenko, I. P. (1967) Promyshlennoye vinogradarstvo na Terskikh peskakh (Commercial grape growing on the Terek sands), *V.S.N.* 1967 (6), 45–50.

Lovell, C. A. K. (1969) The role of private subsidiary farming during the Soviet Seven-Year Plan, 1959–65, *Soviet Studies*, 20, 46–66.

Loza, G. M. (1967) Sistemy vedeniya sel'skogo khozyaystva v razlichnykh prirodno-ekonomicheskikh zonakh, (Systems of agricultural management in different natural-economic zones), *V.S.N.* 1967 (5) 75–83.

Lyashchenko, P. I. (1947–56) *Istoriya narodnogo khozyaystva SSSR*, 3 vols. Leningrad, first edition (1939) trans. by L. M. Herman, *History of the national economy of Russia to the 1917 revolution*, New York, Macmillan, 1949.

Lydolph, P. E. (1964) The Russian sukhovey, *A.A.A.G.* 54, 291–309.

Markova, K. V. and others (1967) Perspectivy povysheniya belkovosti moloka (Prospects for the increase of protein content in milk), *V.S.N.* 1967 (7), 59–65.

Maynard, J. (1942) *The Russian peasant and other studies*, London, Gollancz.

Meisel, J. H. and Kozera, E. S. (1953) *Materials for the study of the Soviet system*, 2nd. ed., Ann Arbor, George Wahr.

Mellor, R. E. H. (1964) *Geography of the U.S.S.R.*, London, Macmillan.

Miller, M. (1926) *The economic development of Russia 1905–14*, 2nd. ed. 1967, London, Cass.

Milovanov, V. K. (ed.) and others (1960) *Al'bom po iskusstvennomu osemeneniya sel'skokhozyaystvennykh zhivotnykh*, Moscow, trans. by A. Birron and Z. S. Cole, *Artificial insemination of livestock in the U.S.S.R.*, Jerusalem, I.P.S.T., 1964.

Morozov, V. I. (1967) Osvoyeniye zakurstarennykh zemel'metodom frezerovaniya (The reclamation of scrub-grown areas by rotary cultivation) *V.S.N.* 1967 (2), 46–51.

Mosse, W. E. Stolypin's villages, *Slavonic and East European Review*, 43, 257–274.

Narkiewicz, O. A. (1966) Stalin, war communism and collectivization, *Soviet Studies*, 18, 20–37.

Narkiewicz, O. A. (1968) Soviet administration and the grain crisis of 1927–28, *Soviet Studies*, 20, 235–241.

Nemchinov, V. (1955) Voprosy spetsializatsii proizvodstva pri perspectivnom razmeshchenii sel'skogo khozyaystva (Questions of the specialisation of production with the perspective of agricultural distribution), *Planovoye Khozyaystva*, 1955 (4), 60–61.

Nemchinov, V. (1956) Economic questions in the development of animal breeding, in Polyakova (ed.) 67–92.

Nikishov, M. I. (1960) Experience in distinguishing agricultural zones and regions on the agricultural map of the U.S.S.R., *S.G.* 1 (10), 23–32.

Nikitin, N. P., Prozorov, E. D. and Tutykhin, B. A. (1966) *Ekonomicheskaya geografiya SSSR*, Moscow.

Nimitz, N. (1967) Farm employment in the Soviet Union, 1928–1963, in Karcz (ed.), 175–211.

Nove, A. (1969) *Economic history of the U.S.S.R.*, London, Allen Lane.

Novikov, I. (1956) Lowering production costs and securing profitability in state farms, in Polyakova (ed.), 284–296.

Novikov, E. A., Startsev, D. I. and Arzumanyan, E. A. (1950) *Plamennoye delo v skotovodstve*, Moscow, trans. by A. Birron, R. Farkash and M. Paenson, Breed improvement in cattle breeding, I.P.S.T. ,Washington, 1960.

Pakhtusov, Z. I. (1967) O povyshenii zhirnomolochnosti krupnogo rogatogo skota (On increasing butterfat content in cows' milk), *V.S.N.*, 1967 (2), 68–74.

Pannikov, V. D., Alekseyev, P. F. and others (1967) *Sel'skoye khozyaystvo SSSR* (Agriculture of the U.S.S.R.), Moscow.

Parker, W. H. (1968) *An historical geography of Russia*, London, Univ. of London Press.

Pavlov, A. A. and Gaabe, Yu. E. (1964) *Statistika sel'skogo khozyaystva*, Moscow.

Pavlovsky, G. (1930), *Agricultural Russia on the eve of the Revolution*, London, Routledge.

Pereverzev, V. N. and Golovko, E. A. (1968) Effect of cultivation on the physicochemical properties and biological activity of peat-bog soils, *Soviet Soil Science* (*Pochvovedeniye*) 1968 (3) 359–367.

Ploss, S. I. (1965) *Conflict and decision-making in Soviet Russia: a case study of agricultural policy 1953–1963*, Princeton, Princeton Univ. Press.

Polyakova, N. (ed.) (1956) *Voprosy ekonomiki sel'skogo khozyaystva*, Moscow, trans. by A. Farkash, *Problems of agricultural economy*, Washington, I.P.S.T., 1960.

Pospielovsky, D. (1970) The 'link system' in Soviet agriculture, *Soviet Studies*, 21, 411–435.

Rakitnikov, A. N. (1970) *Geografiya sel'skogo khozyaystva* (*problemy i metody issledovaniya*), (Agricultural geography, problems and research methods), Moscow.

Rakitnikov, A. N. and Kryuchkov, V. G. (1965) Agricultural regionalization, *Geografiya SSSR, No. 2, Economic regionalization of the U.S.S.R.*, Moscow, trans. in *S.G.* 7 (5), 1966, 48–58.

Raup, P. M. (1967) Comment, on paper by N. M. Jasny, in Karcz (ed.) 258–264.

Robinson, G. T. (1932) *Rural Russia under the old regime*, New York, Macmillan.

Rosenko, A. K. (1967) Sort, urozhaynost', sebyestoymost i pribyl' (Variety, yield, prime cost and profit), *V.S.N.* 1967 (2) 102–106.

Rostovtsev, N. F. (ed.) 1961 *Teoriya i praktika razvedeniya sel'skokhozyaystvennykh zhivotnykh*, Moscow, trans. by D. Vinograd, *Theory and practice of livestock breeding*, Jerusalem, I.P.S.T., 1966.

Rubinow, I. M. (1908) *Russia's wheat trade*, U.S. Dept. of Agriculture, Bureau of Statistics, Bulletin 65, Washington.

Rybakov, B. A. (ed.) (1966) *Istoriya SSSR* (1st series), Moscow, Akademiya Nauk SSSR.

Samokhvalov, N. F. (1957) The climatic make-up of sukhoveis in Kazakhstan, in Dzerdzeevskiy (ed.) 46–53.

Sapozhnikova, S. A. and Shashko, D. I. (1960) Agroclimatic conditions of the distribution and specialization of agriculture, *S.G.* 1 (9), 20–43.

Sapozhnikova, S. A. and Shashko, D. I. (1962) *Agroklimaticheskiye resursy SSSR; voprosy razmeshcheniya i spetsializatsii sel'skogo khozyaystva SSSR* (Agroclimatic resources of the USSR; questions on the distribution and specialization of agriculture in the U.S.S.R.,) Moscow.

Sauer, C. O. (1952) *Agricultural origins and dispersals*, New York, American Geographical Society.

Saushkin, Yu. G., Nikol'skiy, I. V. and Korovitsyn, V. P. (eds.) (1967) *Ekonomicheskaya geografiya Sovetskogo Soyuza*, Moscow.

Schwartz, H. (1965) *The Soviet economy since Stalin*, Philadelphia and New York, Lippincott.

Selyaninov, G. T. (1957) The interpretation of drought and sukhovei in agronomy and their distribution in the European territory of the Soviet Union, in Dzerdzeevskiy (ed.) 14–22.

Senchenko, G. (1967) Konoplya, kenaf, dzhut (Cotton, hemp, jute) in Pannikov and others, 119–120.

Seton-Watson, H. (1952) *The decline of Imperial Russia*, London, Methuen.

Shashko, D. I. (1962) Climate resources of Soviet agriculture, in Akademiya Nauk SSSR *Soil-geographical zoning of the U.S.S.R.*, 378–445.

Sheehy, A. (1967) Irrigation in the Amu-Dar'ya basin: progress report, *Central Asian Review*, 15 (4), 342–353.

Shimkin, D. B. (1963) Current characteristics and problems of the Soviet rural population, in Laird (ed.), 79–127.

Shishkin, N. I. (1961) *Trudovyye resursy SSSR* (Labour resources of the U.S.S.R.). Moscow.

Sidorov, A. P. (1958) Izderzhki proizvodstva pri razlichnykh vidakh transporta i transportnaya set' Novosibirskoy oblasti (Production and transport costs in Novosibirsk oblast) in Sinyagin (ed.) vol. 2, 409–431.

Sil'vestrov, S. I. (1965) Geograficheskiye osnovy bor'by s eroziyey pochv (The geographical basis of the fight with soil erosion) *Izvestiya Akademiy Nauk SSSR Seriya Geograficheskaya*, 1965 (2) 49–58.

Sinyagin, I. I. (ed.) (1958) *O sisteme vedeniya sel'skogo khozyaystva v Novosibirskoy oblasti*, (On the system of management of agriculture in Novosibirsk oblast), 2 vols., Novosibirsk.

Skachkov, I. A. (1967) Rol' sel'skokhozyaystvennoy nauki v zashchite pochv ot erozii (The role of agricultural science in erosion control), *V.S.N.* 1967 (7), 20–23.

Skoropanov, S. G. (1968) Some conclusions derived from the practice of melioration of peat-bog soils, *Soviet Soil Science* (*Pochvovedeniye*), 1968 (1), 1–6.

Skvortsov, A. A. (1949) Svoystva i vlagoobespechennost' yestestvennykh isparyayushchikh poverkhnostey (Properties and water content of natural evaporating surfaces), *Meteorologiya i Gidrologiya*, 1949 (4).

Slavkov, M. I. and Tyutyunnik, M. E. (1968) *Ekonomika aviatsionno-khimicheskikh rabot* (Economics of aviation-chemical work), Moscow.

Smith, E. (1967) Soviet agriculture; the Khrushchev era and afterwards, *Journal of Agricultural Economics*, 8, 387–402.

Smith, R. E. F. (1959) *The origins of farming in Russia*, Paris, Mouton.

Smolin, N. (1953) O zachatkakh produktoobmena (Concerning the embryo product-exchanges), *Voprosy Ekonomiki*, 1953 (1), 33–45.

Sotnikov, V. P. (1960) Farming problems in the zones of the U.S.S.R. and the tasks of Soviet geographers, *S.G.* 1 (10), 3–22.

Stalin, J. (1952) *Economic problems of socialism in the U.S.S.R.*, Moscow.

Stepanov, V. N. (1957) *Biologicheskaya klassifikatsiya sel'skokhozyaystvennykh rasteniy polevoy kul'tury* (Biological classification of agricultural field crops), Moscow.

Storch, H. (1801) *Tableau historique et statistique de L'Empire de Russie à la fin du 18e siecle*, 2 vol., Basle and Paris.

Strauss, E. (1969) *Soviet agriculture in perspective*, London, Allen and Unwin.

Strauss, E. (1970) The Soviet dairy economy, *Soviet Studies*, 21, 269–296.

Sumner, B. H. (1947) *Survey of Russian History*, 2nd ed. London, Duckworth, 1961 ed. Methuen.

Suslov, V. (1967) Samyy maslichnyy (Oil-producing crops) in Pannikov and others, 118–119.

Swearer, H. (1963) Agricultural administration under Khrushchev, in Laird (ed.), 9–40.

Symons, L. (1967) *Agricultural Geography*, London, Bell.

Taylor, J. W. R. (ed.) (1970) *Jane's all the world's aircraft*, London, Sampson Low.

Teryayeva, A. (1968) Zonal'nye problemy vosproizvodstva rabochey sily i oplata truda v sel'skom khozyaystve (Zonal problems of reproduction of labour power and payment of labour in agriculture.) *Voprosy Ekonomiki*, 1968 (11), 41–51.

Thran, P. and Broekhuizen, S. (1965–9) *Agro-ecological atlas of cereal growing in Europe*, 2 vols., Amsterdam, Elsevier.

Tikhomirov, M. I. (1968) (ed.) *Razmeshcheniye i spetsializatsiya zemledeliya i zhivotnovodstva v zone zapadnoy sibiri*, (Distribution and specialisation of cultivation and livestock rearing in the west Siberian zone), Moscow.

Timoshenko, V. P. (1932) *Agricultural Russia and the wheat problem*, Stanford, Stanford Univ. Press.

Treadgold, D. W. (1957) *The great Siberian migration; government and peasant in resettlement from Emancipation to the First World War*, Princeton, Princeton Univ. Press.

Tsuberbiller, E. A. (1957) Agrometeorological criteria for sukhoveis, in Dzerdzeevskiy (ed.) 65–73.

Udachin, S. A. (1967) Sotsialisticheskiye preobrazovaniya v zemlepol'-zovanii (Socialist transformations in land utilisation) *V.S.N.* 1967 (7), 1–7.

Vavilov, N. I. (ed. and principal author) (1935) *Teoreticheskiye osnovy selektsii rasteniy*, Moscow; selected writings translated in 'The origin, variation, immunity and breeding of cultivated plants', *Chronica Botanica*, 13, 1949–50, Waltham, Mass. 1951.

Venzher, V. G. (1960) *Ispol'zovaniye zakona stoimosti v kolkhoznom proizvodstve* (The utilisation of the law of value in collective farm production), Moscow, 2nd ed. 1965.

Venzher, V. G. (1966) *Kolkhoznyy stroy na sovremennom etape* (The collective-farm system at the present stage), Moscow.

Vernadsky, G. (1948) *Kievan Russia*, New Haven, Yale Univ. Press.

Vilenskiy, D. G. (1957) *Pochvovedeniye*, Moscow, trans. by A. Birron and Z. S. Cole, *Soil Science*, Jerusalem, I.P.S.T., 1963.

Vil'yams, V. R. *Sobraniye sochineniy* (Collected works) 1948–53 Moscow.

Vil'yams, V. R. *Izbrannye Sochineniya* (Selected works) Moscow, 1950–55.

Volin, L. (1951) *A survey of Soviet Russian agriculture*, United States Dept. of Agriculture Monograph 5.

Volin, L. (1967) Khrushchev and the Soviet agricultural scene, in Karcz (ed.), 1–21.

Von Laue, T. H. (1961) Russian peasants in the factory, 1892–1904, *Journal of Economic History*, 21, 61–80.

Vsyakikh, A. S. (1961) Theoretical bases of developing new cattle breeds, in Rostovstsev, (ed.) 88–94.

Wädekin, K. E. (1967) *Privatproduzenten in der sowjetischen Landwirtschaft*, Köln, Verlag Wissenschaft und Politik.

Wädekin, K. E. (1969) Manpower in Soviet agriculture—some post-Khrushchev developments, *Soviet Studies*, 20, 281–305.

Walters, H. E. and Judy, R. W. (1967) Soviet agricultural output by 1970, in Karcz (ed.), 306–344.

Westwood, J. N. (1964) *A history of Russian railways*, London, Allen and Unwin.

Whitman, J. T. (1956) The kolkhoz market, *Soviet Studies*, 7, 384–408.

Wilber, C. K. (1969) The role of agriculture in Soviet economic development, *Land Economics*, 45, 87–96.

Williams, E. W. (1962) *Freight transportation in the Soviet Union*, Princeton, Princeton Univ. Press.

Williams, V. R. see Vil'yams, V. R.

Wronski, H. (1957) *Rémunération et niveau de vie dans les kolkhoz: le troudoden*, Paris, Société d'Edition d'Enseignement Supérior.

Yevseyev, P. K. see Evseev, P. K.

Zal'tsman, L. (1956) The new method of agricultural planning and questions on the zonal distribution of livestock raising, in Polyakova (ed.), 227–242.

Zelenin, I. E. (1966) *Zernovye sovkhozy SSSR* (*1933–1941*), (Grain sovkhozes of the U.S.S.R. 1931–1941), Moscow.

Zhukovskiy, N. I. (1958) Sovremennoye sostoyaniye i perspektivy razvitiya sel'skogo khozyaystva Novosibirskoy oblasti (Present condition and prospects of development of agriculture in Novosibirsk oblast), in Sinyagin (ed.) 14–44.

Zoerb, C. (1964) The Virgin Land Territory: plans, performance, prospects, *Studies on the Soviet Union*, 3 (4), 29–44.

Zvorykin, K. V. (1965) The role of geographers in work on a land cadaster and on an agricultural evaluation of lands, *Voprosy Geografii*, no. 67, 11–23, trans. in *S.G.* 9 (3), 1968, 162–171.

Index

Page references to the meaning of Russian terms will be found under the appropriate entry in this index.